MW01701282

Practi[cal]
Strategies
for Successful
Classrooms

Helping English Language Learners Succeed

Carmen Zuñiga Dunlap, Ph.D.
and Evelyn Marino Weisman, Ph.D.

Shell Educational Publishing

Helping English Language Learners Succeed

Editor
Maria Elvira Gallardo, M.A.

Curriculum Project Manager
Maria Elvira Gallardo, M.A.

Editor-in-Chief
Sharon Coan, M.S. Ed.

Creative Director
Lee Aucoin

Cover Design
Lee Aucoin
Lesley Palmer

Imaging
Alfred Lau

Production Manager
Phil Garcia

Publisher
Corinne Burton, M.A. Ed.

Shell Educational Publishing
5301 Oceanus Drive
Huntington Beach, CA 92649-1030
www.seppub.com

ISBN-0-7439-9109-5

Table of Contents

Table of Contents *(cont.)*

Introduction

The year is 1990. You are a new teacher in Minneapolis. 80% of the students in your classroom are native English-speaking Anglo-Americans.

Fast forward to the 2003–04 school year. The same classroom in the same school in the same neighborhood now looks like this—four of your students are native Spanish speakers, three are native speakers of an Asian language, one is American Indian, and the remaining 20 of your students are divided between 12 African-American students and eight Anglo-American students (Minneapolis Public Schools, 2005).

The demographic changes illustrated above are a reflection of how the English learner population in the United States has grown in the last decade. In Los Angeles, Chicago, Boston, New York, and Miami, and in fact in all regions of the United States, the English learner population has mushroomed. In the 2001–02 academic year, 4.7 million school-age students, or 9.8%, were English learners (Nieto, 2004). Yet despite this ever-growing English learner student population, teachers report that they have very little professional preparation—not sufficient to feel prepared to successfully teach English learners (Meskill, 2005; Gandara, Maxwell-Jolly & Driscoll, 2005).

Preparing teachers to find effective ways to work with English learners has become a critical element of good teaching. As a new teacher, whether you have just a few

English learners or a full class, you'll want to develop effective and appropriate ways to meet their learning needs. This book is for you.

In the chapters that lie ahead, we present practical advice and ideas, weaving in theory and research to enrich your understanding of teaching English learners. From our many years of K–8 teaching and university instruction in teacher preparation programs, we have gathered a body of experience and information. Based on these, we are pleased to offer you—a new teacher—suggestions and ideas that we wish we had known when we began our careers as teachers in classrooms with English learners.

Please use this book as a resource. Ideally, you will use it in conjunction with a good reading/language arts methods textbook. This book is not meant to be an English as a Second Language program. It is intended to provide background information, underlying principles, and ideas for you to put to use in your classroom as you work with English learners.

How this book is organized

In each chapter you will find an overview of the main topic with specific connections to English learners and post-reading reflection questions to reflect on the material.

Chapter One gives you an overview of the key elements of language learning and reasons that each element is useful for you to understand as a classroom teacher. **Chapter Two** looks at the many influences on English learners both in and out of school. **Chapter Three** offers you some assessment tools for use in your classroom. **Chapters Four, Five**, and **Six** address oral language, reading, and writing development. **Chapter Seven** looks at content instruction.

Chapters Four, Five, Six, and **Seven** also include the following elements:

- specific instructional strategies and approaches—or what we call *tools*—for your classroom use
- classroom vignettes with teachers who put some of these tools to use; that is, you'll read about good practice in action
- an opportunity for you to identify the tools that teachers use to assist their English learners
- space for you to reflect on what you have read and learned in each chapter

Chapter One

Understanding Language

An effective teacher of English learners understands basic elements of language and language development. This is important for two reasons. First, you will actually be able to observe these aspects of language in your students. Second, this information will give you insight and help you understand ways to guide your English learners. At the end of each element of language, you will read why this knowledge is useful for you as a classroom teacher. Please understand that volumes have been written on any single aspect of language that you read about below, should you be interested in exploring any one topic. What you read here are the briefest of explanations. Before you read the chapter, think about your answers to the following questions: *How do you define language? What influences how we use language?*

Eight Elements of Language

What is language? How can it be defined? Like the air around us, it's everywhere, and so much a part of our environment that we may not give it much thought. Minimally, language is a series of arbitrary sounds strung together that permit a group of people to communicate.

However, it is much more than this—it is a universal human phenomenon that is the foundation of all our communication. It is systematic and rule-governed, influenced by culture, social and economic class, and even when, where, and with whom we use it. Let's take a closer look.

Language . . .

1. develops naturally
2. develops in stages
3. has structure
4. is intertwined with culture
5. is linked to cognition
6. has varieties
7. is learned in social contexts
8. is influenced by purpose and context

1. Language develops naturally

Unlike any other learned phenomenon, humans acquire their first language largely by hearing it and by interacting with speakers in their environment. The same cannot be said about learning to play the piano, learning to ride a bike, learning to write, or any other learned behavior. Three basic theoretical approaches to language learning provide a different lens on the process. Linguistic theory holds that language has a structure that is unique and distinct, and that babies are born with specific language learning mechanisms that enable them to learn language in a relatively short period of time. Chomsky's work (1965) provided the initial theoretical support for this view. Cognitive theories of language learning state that it is directly linked to stages of cognitive development. We associate the work of Piaget and Bruner with this theoretical lens. Social interactionists, as the label indicates, believe that the key element of language learning resides in meaningful social interactions within a supportive environment. Vygotsky (1978) discussed the critical role of interacting with others in a stimulating environment.

Certainly, there are intuitively appealing aspects of each approach. A stance that blends elements of each is something like this: babies are born with an innate language learning ability that differs from other types of abilities. This enables them to quickly grasp the structures of language. Furthermore, language and cognitive development influence each other and, in order for language to fully blossom, social interaction is necessary. Regardless of which theoretical stance of the language learning process one may take, we know that it is a human and naturally occurring phenomenon that develops in strikingly similar ways across cultures, languages, and geographical locations.

Why is this useful information for a teacher?

Teachers can create learning environments for English learners that capitalize on their innate ability to learn language. This is particularly true for younger children who are still in the later stages of natural, first language acquisition—between the ages of four and six. Older students can benefit by being reminded that just as they learned their first language through listening and through general exposure to language, they will benefit from actively seeking exposure and social interaction with others who can provide meaningful input in the second language. Furthermore, they—and you, the teacher—can enhance students' English language skills by placing language learning in meaningful and interesting contexts. We will return to this point later.

2. Language develops in stages

Every living thing passes through stages of development—whether an embryo or a new teacher. Similarly, language development also occurs in stages. In the first few months of life, babies don't vocalize much beyond crying, gurgling, cooing, or other minimal random sounds. However, they are taking in enormous amounts of auditory information. This early period provides the foundation for later language development. Within a few short months, babies begin babbling. Linguists suggest that these sounds provide a baby with practice for tongue and mouth movements in preparation for pronouncing words. These precursors to language turn into recognizable distinct words around the age of 12 months, the time when a joyous parent hears the first distinguishable word. Other new words follow in quick succession, usually between ages three and five.

In these few sentences, we have described three distinct stages of language development—a silent period, babbling, and first words. While they are common stages that occur universally at about the same age, there is some individual variation. We will return to the matter of stages of oral language development in Chapter Four.

Why is this useful information for a teacher?

You will clearly see your students passing through various stages of English language development. Furthermore, you will see that each student has his/her own pace of developing English, in the same way that babies and young children have in their native language. Remember, while you can foster language development, you cannot force it. You can provide a rich language environment for a young child and engage in multiple opportunities for conversational give-and-take. However, you cannot force a 12-month-old to pronounce "daddy" instead of "dada" or an 18-month-old to understand a lengthy sentence. Similarly, while there is much you can do to encourage and support English language development in your students, recognize that an individual student's rate of language development runs by an internal clock. You cannot rush Mother Nature!

3. Language has structure

The flow of language can be compared to music. Imagine listening to a beautiful piece of music, or listening to someone speak who "has a way with words." We don't usually think about the underlying structure that makes up this pleasing "whole." However, it is precisely because the various elements of the structure are placed together in just the right way that creates their appeal. Two elements that comprise musical structure include rhythm and the particular key in which the melody is written. Let's take a look at the structure of language.

Phonology is the study of sounds. The smallest unit of sound is a *phoneme*. An example of a phoneme is "n"—or /n/, as a linguist writes it. There are rules in English, as there are in every language, as to how phonemes may be arranged to make words. In English, one rule allows us to place /n/ and /d/ at the end of a word—as in *stand*—but not at the beginning of a word. In Swahili, a phonological rule permits /n/ and /d/ at the beginning of a word—as in *ndiyo*, meaning *yes* or *it is so*. Stress, pitch, and tone also affect the way we produce sounds. Stress falls on a word within a group of words or on a syllable within a single word. Pitch and tone affect a single sound. In English, changing the pitch and tone of a sound do not cause meaning change. For example, English speakers would agree that the meaning of the word *ma* is the same in each of the following sentences, regardless of pitch and tone: *My ma was born in December. Ma, come quick!* In tonal languages, pitch and tone create completely different meanings. In Vietnamese, for example, *ma* can be said in five different ways—with rising, falling, or level pitch and tone, or a combination of these. Each of these creates completely different words—*mother, ghost, gravestone, horse*, and *a type of chemical reaction*.

Morphology looks at the structure of meaningful units of sound. A *morpheme* is the smallest unit of meaning. There are *free* and *bound* morphemes. A free morpheme carries its own meaning, like *hat* or *desk*. Bound morphemes consist of prefixes, suffixes, and inflected endings such as *–ed* or *–ing* and must be combined with other words. Inflected morphemes are grammatical markers. Young children and English learners begin to use and manipulate inflected morphemes in useful and sometimes curious ways as they become proficient. Consider the morpheme *–y*, meaning roughly "full of," as in *rusty* or *dusty*. A three-year-old was recently overheard referring to someone as *singy*, meaning "someone who is full of song or who likes to sing". Morphological development and use indicates that a finer-grained understanding of language is in process.

Syntax is the system of rules that govern how words are arranged to form meaningful phrases and sentences. Syntax accounts for the word order, or *linear order*, of a sentence or a phrase and the meaningful groupings of words called *constituents*. Syntactic rules also explain how *ambiguous* and *paraphrased* sentences are related. This is accomplished by describing their *underlying structure*. An example of an ambiguous sentence is: *The freeway sign read "California left," so Ed turned around and drove home.* Here are three examples of paraphrased sentences: *John thinks he's intelligent. John thinks himself intelligent. Why does John think he's intelligent?* Finally, syntax describes how a sentence can be expanded, or be *recursive*, which is the term linguists use. Here are a few sentences that are recursive, or expanded: *Terry put on his hat. Terry put on his blue hat and tan jacket. Terry quickly put on his blue hat and tan jacket, and ran* *out the door.* A complete syntax should account for the creation of the infinite number of sentences possible in a language.

Semantics is the study of the meaning of words, phrases, and sentences. Describing meaning is a much more fluid proposition than describing rules that govern other aspects of language. Indeed, it is the most elusive aspect of language to describe in a systematic way. This is because meaning is influenced by use in context and by individual and cultural aspects. Linguists have constructed a rather complex set of categories and principles that describe some aspects of semantics. For example, the language philosopher John Langshaw Austin (1962) described what he called "felicity conditions," meaning that a sentence can be taken seriously only under a suitable circumstance. The sentence *I now declare you husband and wife* can be taken seriously only if spoken by an authorized person within the appropriate circumstance. Partial descriptions make the study of semantics still very much open to discussion and study.

Pragmatics looks at how language is used for real-time on-going communication. Linguists have developed principles and guidelines that describe pragmatics. One of these is called *speech acts* (Austen, 1962). These describe how we do things and get others to do things with words. We invite, command, pardon, apologize, and a host of other acts. *How* we do things with words is culturally embedded. How we invite, decline an invitation, greet, and take leave are directly linked to culture. For example, in Japan, it is expected that someone will turn down an invitation by simply saying "I am not able to come." In the U.S., we expect either a bit of an apology or an explanation. Instruction for English learners should include attention to the use of pragmatics, or how native speakers use English in social contexts.

These five structures of language—phonology, morphology, syntax, semantics, and pragmatics—are universal.

How they interact and the importance and prominence of particular elements are language-specific.

Why is this useful information for a teacher?

Your English learners bring a vast amount of knowledge about language structure in their first language to the process of learning English. They rely on familiar linguistic rules that govern their first language to help them learn and navigate English. Linguists refer to this as *transfer*. Your students' knowledge and reliance on the first language can be very helpful, particularly for older English learners. For example, an older student will know there are ways of expressing concepts such as past tense, relationships, time, and so on. This general knowledge helps the older learner to specifically seek ways to express these concepts in English. Sometimes, however, influence from the first language may cause difficulty, or *interference,* with English. For example, if adding the morpheme *–s* or *–es* to make a noun plural is not a rule in the first language, it will be difficult for the student to internalize and use this rule in English. Knowing something about the structure of a student's native language can help you predict specifically challenging aspects he/she may have in learning English.

4. Language is intertwined with culture.

Discussing the link between language and culture with university students and inviting them to share personal experiences calls forward many amusing stories of what can happen when there is a disconnect between the two. Knowing a language does not necessarily mean that one can navigate one's way in the culture of that language. For example, perhaps your American students have slumber parties or sleepovers. In the United States, this is a common way for children to "hang out" at a friend's house and have fun into the evening and often beyond. For people of other cultures, this may be a bizarre and uncomfortable concept—allowing one's children to sleep in the home of an another person. So not only does the phrase "slumber party" require an explanation, it also requires becoming familiar with a common North American concept that is an integral part of the culture.

Why is this useful information for a teacher?

You will find it necessary to not only teach English, but to sometimes also help students navigate cultural differences as they become acclimated to new linguistic meanings and cultural traditions or events.

5. Language is linked to cognition

We stated above that some theorists believe that cognitive development supports language development. Some believe the reverse is true—that language development drives learning and provides humans the ability to organize their thinking. Certainly, there is an undeniable link between the two. Let's look at two examples. Consider a young child who makes the conceptual discovery of *in* and links it to the word. Suddenly he/she begins to correctly use this cognitive and linguistic discovery incessantly—"in the box," "in the shoe," "in the bed," "in the doggie," and so on. Children delight in making these discoveries. The joy is evident in the constant use they make of their exciting new finds.

Conversely, here is an example illustrating how language can influence understanding. One of our children at around age three became intrigued with the word *area*. She used the word constantly, as in "the silverware area" (the drawers where the silverware was kept) "the desk area" (in reference to her little chair and table with her crayons, paper, and books), and so on. Through hearing the word used by others, she began to refine her understanding of it until she came to use the word correctly. These examples illustrate how cognition affects language and, conversely, how language affects cognition.

Why is this useful information for a teacher?

A teacher's most important responsibility is helping students learn content and develop the vocabulary to understand, speak, read, and write about what they learn. Vocabulary development is one of the most challenging aspects of learning another language. Vocabulary and concepts are directly linked. It is imperative that teachers of English learners embed learning and vocabulary in understandable contexts. Chapter Five discusses ways to promote vocabulary development. Chapter Seven discusses ways to help students understand concepts in the content areas.

6. Language has varieties

We can think of language variety in two different ways. First, variety can refer to the scope of sounds and sound systems that are humanly possible. Some languages spoken in southern Africa incorporate different types of clicks. There are languages that include nasal sounds, throaty guttural sounds, and variations on tone and pitch of a single sound. Humans have included an amazing array of sounds found in languages around the world. A second way we can think about language variety is within a single language. Spoken English varies across regions and even socioeconomic class. Every language has such varieties. A regional variety is usually referred to as an *accent* and implies a difference in pronunciation. A regional variety may also be called a *dialect.* Note, however, that dialects may include grammatical and lexical (or word) variations, as well as differences in pronunciation. An example of a grammatical variation that is not standard English is *I* ***so*** *do not want to go* (*so* is stressed and used as an adverb rather than an adjective). The use of *so* in this instance has become common among certain groups, yet it is not considered "standard" English. The words *bag, sack,* and *poke* are examples of regional lexical variations. They mean the same thing; however, they are used preferentially in different regions of the United States. Individuals typically don't think they have an accent or speak in a certain dialect because they are surrounded by others who speak the same way. We consider our own speech the norm.

Why is this useful information for a teacher?

English learners will have difficulty hearing and pronouncing sounds in English that do not exist in their native languages. Young students who are learning English while learning to read in English can be greatly challenged as they develop phonemic awareness and learn phonics rules for decoding text. Depending on a student's primary language, some areas of difficulty for English learners are *b/v*, *l/r*, and *p/f* distinctions and developing the ability to distinguish and pronounce the large array of English vowel sounds. Helping English learners develop phonemic awareness and then helping them learn to read sounds as represented by written symbols is a critical issue that will be discussed in Chapter Five. Varieties of spoken English, however, have little, if any, influence on English learners. Only if the spoken variety differs greatly in syntactic structure from standard written English might it present a challenge to the English learner.

7. Language is learned in social contexts

We stated above that social interactionists believe that language learning resides in meaningful social interactions within a supportive environment. Humans are social and curious, and we want to be included when we're interested in what's happening around us. So, an environment that sparks a desire to be involved and to know what is going on is critical. Another important element for the learner is a guide who knows how to listen, how to explain, how to answer questions, and how to negotiate meaning in that environment. For young children, caretakers are the guides who interact in meaningful ways. For English learners, the teacher and friends are the guides who construct meaning through comprehensible language input. Of course, the learner is not a passive recipient. Negotiating meaning is a joint venture. Learner and guide are partners in constructing meaning. A good guide modifies and adjusts language to the level of the learner for maximum understanding as the learner participates in attempting to understand.

Why is this useful information for a teacher?

As the teacher, your role is critical in providing a rich context in which your students can engage in learning and, consequently, learn English. With this in mind, you can do a great deal to set up your classroom environment so that students have multiple opportunities to talk to one another as they explain, clarify, complete projects, and construct meaning together. You can also model for your native English speakers how to be helpful guides for the English learners in the classroom. In Chapters Two and Seven, we will talk more about providing rich social contexts for English learners.

8. Language is influenced by purpose and context

Maybe you have noticed that the way you use language differs depending on the person you are talking to and maybe even the context of the conversation. For example, you use language differently with your dentist versus how you speak with your students. You speak with close family members in a casual manner, which is different from conversing with someone you meet for the first time in a formal setting. Sociolinguists have even studied the different ways in which men and women use language. Classroom language varies as well. An important distinction is language that is used for social purposes versus academic purposes. Social language is *here-and-now* language, whose meaning is context-linked and obvious. Playground or lunchtime language are examples of social language. Similarly, written language depends on the purpose and audience. A note you write to a friend is far more casual than a paper you write for a class. Talking about academic concepts and writing formally is vastly different from informal uses of spoken and written language.

Why is this useful information for a teacher?

Social language will be more easily learned and used by your English learners, while academic language will require more effort and direct focus. It is important for you to fully understand the differences between social and academic language so that you can help your English learners in the best ways possible. Cummins (1981) has discussed these types of classroom language use. We will draw on his work in Chapter Two to elaborate on these uses of language. Throughout the book, we discuss strategies and approaches—we call them *tools*—that will help you scaffold meaningful academic language for your students as they negotiate to understand and learn English.

Important Points to Remember

- Everyone passes through similar stages of language development.
- English learners bring a great deal of information from their first language to the process of learning English.
- Language learning relies heavily on social and cognitive support.
- Learning English can be supported by the environment and helpful individuals; however, learning a new language takes time.

Post-Reading Reflection

1. What was your definition of language at the beginning of the chapter?

2. Would you revise your initial definition? If so, how?

3. Reflect on three elements of language and why these are important for teachers to understand.

Chapter Two

Understanding Your Students Both In and Outside of Your Classroom

As a teacher of English learners, you have a powerful influence on your students' learning, as do the classroom and school environments. Beyond the school, family, community, and culture also exert influences. It is especially important that you understand your English learner not just in the context of your classroom, but also in the realm of broader influences. In this chapter, we explore these influences that will help you understand your students better, provide a rich learning environment for them, and set the stage for success in your classroom and beyond.

Figure 2.1: Influences on your English Learners

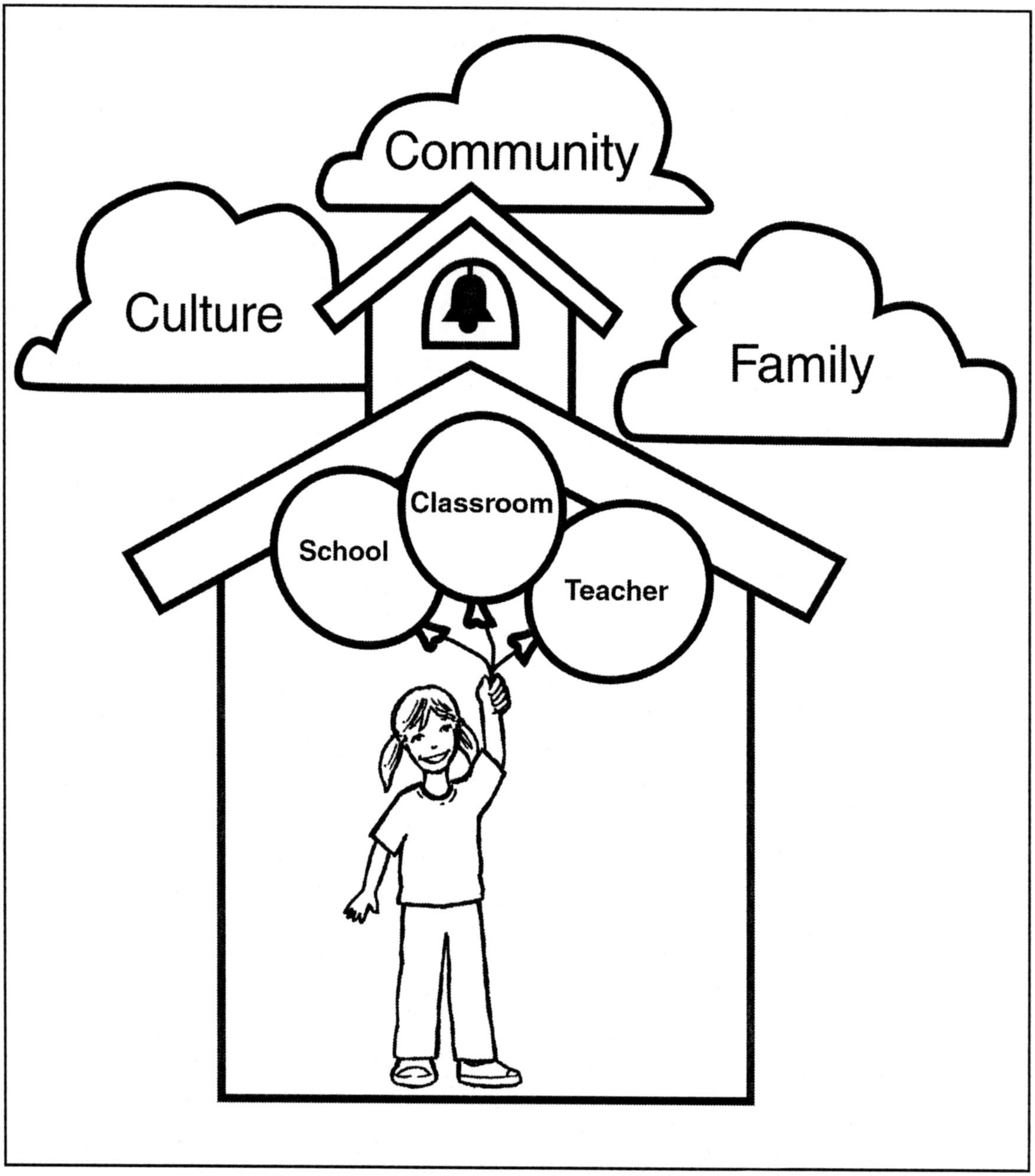

The Culture of Schooling

Carmen's daughter attended school in Brazil when the family lived there. Carmen found it difficult—indeed, impossible—to let go of her beliefs about what constitutes good teaching. One day she gathered her daughter and friends during after-school hours, read to them, and invited the children to talk about their favorite part or act out portions of the story. The children looked very confused. They were unable to do this. Why not? It came as a great surprise when she found that children from an early school age were expected to attend carefully to teachers, memorize a great deal of material, and respond by rote memory. Her daughter's friends were unaccustomed to responding to stories, being asked their ideas about what they read, or reflecting on text. Carmen's daughter occasionally found herself in trouble at school precisely because she would share uninvited opinions, which was the norm in her home but not at school. When Carmen discussed instructional approaches with teachers at the school, they were quite certain their lecture and recitation approach was the best.

Culture affects our values and beliefs in ways that we may not even think about. This event illustrates how students become socialized to a culture's way of schooling. It also illustrates that views of what constitutes good teaching vary from country to country. How a culture does schooling is unique unto itself.

Expectations concerning the role of the teacher also vary from culture to culture. For example, in Vietnam, the teacher is revered for the knowledge he or she has and is held in high esteem. It would be completely inappropriate for a teacher to respond to a student's question by saying, "Gee, I don't have an answer for that but let's look it up and find out." A teacher is simply expected to know everything.

A third aspect that is affected by culture and by the family's socioeconomic status is parent and family involvement in the school. Should parents be involved guides? Quietly supportive? Firm advocates for the perceived needs of the child? Hands-off because they believe expertise resides in the school and it's best to leave instruction to the teachers? In the United States, we expect parent involvement and seek ways to involve them. It is important to understand that your expectations may not match the cultural norm for your student's family. You may need to work at helping parents become comfortable with participating in their child's schooling.

As you work with parents to help them become familiar with how schooling is done in the United States, explain the grading system, test scores, and what they mean. Make personal contact, because letters home—even in the home language—may not be sufficient for establishing a positive relationship and good communication. Hallway conversations, phone calls home with the help of translators, or even home visits are very valuable. If your school (or district) does not offer assistance in a particular language, seek creative ways to look for community resources. Evelyn once called a Romanian church to inquire about bilingual community resources to help with a new student and his family. The church obliged them by offering to send someone to the school right away.

As you find yourself teaching English learners, it is helpful to find out as much as you can about how schooling is done in the native countries of your students. Having advance information will help you know what to expect and enable you to provide a compare-and-contrast for the parents. They may not feel completely satisfied with your teaching approach, but they will appreciate your efforts to explain differences. Furthermore, you and your colleagues may find ways to help them grow to accept the way we do schooling in the United States.

Here are some dimensions to consider that will give you insight to a culture's view of schooling.

1. Is schooling considered competitive or cooperative?
2. Is individual work valued over group work?
3. Does teaching methodology favor student inquiry or lecture and recitation?
4. Is sharing work among students considered cheating or the norm?
5. Are boys and girls segregated or integrated?
6. Are educational expectations and attainment gender-driven?
7. Is student improvement over time satisfactory or is the focus on perfection, excellence, and high grades?
8. Are students expected to focus exclusively on academics or are other interests, such as sports or music, encouraged as part of the curriculum?
9. Do parents and students expect that high-stakes tests are a normal part of schooling or are high-stakes tests unexpected or not understood?
10. Are parents accustomed to being involved in school and their childrens' activities?

Community Influences

In addition to cultural considerations, you must also look to the broader community for language influence. In neighborhoods and communities with a single and large non-English speaking population, there may be little opportunity, reason, or motivation to use English outside the school. In these neighborhoods, most aspects of daily life can easily occur in a language other than English. Conversely, in a multilingual community, English becomes the common language.

In schools and neighborhoods with a large concentration of a single language other than English, you will need to become creative about finding ways to promote and expand ways for your students to use English. This is important because you, the teacher, are the primary English language model. English learners should have multiple English language models and multiple opportunities to use English. Using e-mail to connect native English speakers and English learners—that is, "e-pen pals"—is ideal. Student-to-student connections:

- foster real and meaningful communication;
- provide fluent English language models in addition to the teacher;
- create a real purpose for writing as clearly and correctly as possible; and
- can be used for social purposes and even structured for subject matter exchanges.

We've considered some cultural and community aspects of schooling that affect English learners. Now let's look more closely at family and individual influences that affect a student's ability to learn English.

Family Influences

Family expectations and socioeconomic status play a large role in your students' English development. Typically, the more education the parents have, the more they will encourage and directly support their children in learning English and doing well in school. Keep in mind, however, that many immigrant parents with limited education are eager for their children to learn English and obtain a good education in hopes of bettering their lives. Learning and literacy are generally linked to the level of education of the parents, but not necessarily to the parents' current line of work.

Consider this example from Carmen's teaching experience:

Antonio busses dishes in a restaurant. At a parent conference, it is evident that he values education and is eager for his seventh-grade son to learn English and be successful in school. A bit of polite probing reveals that Antonio graduated from a post-secondary technological institute in Mexico and taught electronics before he and his family came to the United States. Eusebio's lack of English is one factor that contributes to his current job, as well as to his frustration at not finding employment more appropriate for his level of education.

We know of many instances in which parents have achieved significant levels of education in their home countries but are unable to find appropriate work in the United States, so they are forced to accept any type of employment. Assume that your students' parents are eager for them to learn English and do well in school, and that they may or may not demonstrate this support in ways that are immediately obvious to you.

English versus the native language at home

Your student may be the first in the family to learn English, or may have siblings or other family members who bring English into the home environment and who have paved the way in the educational system. Certainly, having other English speakers in the home is helpful. Sometimes preservice teachers ask us how they can encourage the parents of their students to use English at home. Our response is: *don't*. If the parents' dominant language is not English, then they should speak, and if possible, read with their children in that native language. The parents should be encouraged to use the dominant language to the *fullest extent possible* at home. Why? Solid development in any language is far preferable to impoverished language development in English.

Parents should tell and read stories, recite rhymes, sing songs, even watch television together and then discuss what they watched—all in the native language. Parents should talk with their children in the clearest and richest way possible and encourage their children to talk with them in the native language. This use of language helps with concept and vocabulary development. It is in using language to its fullest that helps children develop thinking and the language that goes along with it. Your job as a teacher is to teach English and content, and that job is so much easier when students have a solid grasp of the first language. It is far better to transfer knowledge from one language to another rather than to have to develop concepts in a language that you're learning. Encourage the use of the parents' dominant language at home as much as possible! An added benefit is that the parents and children can stay linguistically connected. Learning English should not mean losing the home language. One of the most unfortunate aspects of losing the home language is that the student may also lose the ability to communicate with his/her family members.

Individual Factors that Affect Learning English

Of course, many factors affect an individual's ability to learn English. The most important ones are age of acquisition, amount of education in the student's primary language, motivation to learn English, and language learning aptitude.

Age of acquisition

Children under the age of about eight learn language differently from older learners. Younger children use innate language learning abilities and can acquire native-like fluency, including pronunciation. Older learners rely on general cognitive learning strategies. They can learn to speak, read, and write perfectly well; however, after about the age of 13, they will begin to lose the ability to completely acquire standard English pronunciation (Lightbown & Spada, 1999).

Education in the primary language

Schooling and literacy level in the primary language influence a student's ability to learn English. Take the case of a high school student from Mexico who had a strong background in the sciences and spoke very little English. When she arrived in the United States, she was placed in a tenth grade biology class that covered material she was familiar with. She earned a B in the course. *How?* She was not learning concepts, but rather how to express these concepts in English. A strong primary language background is a key element in being able to learn English, because the student is not struggling to accomplish two tasks at the same time—learning English and learning *in* English; rather, the student can focus on the single task of learning English. The underlying concept of bilingual education rests on developing strong primary language and literacy skills that the student can then apply to another language. The student whose example we used here experienced what Krashen (1996) refers to as *de facto bilingual education*—a solid primary language education in her home country, followed by education in English. She did not have to learn concepts all over again; rather, she learned how to express what she already knew in English.

One of the most challenging teaching assignments is working with older English learners, say 12 years or older, who have had little or no formal education. These students are burdened with the two tasks mentioned above—learning English and learning *in* English. Here are some suggestions to help you and your students in this type of situation:

- Provide survival English—basic fundamental phrases that will help the students navigate their environment in school and in the community.
- Teach students how to read and write the basic phrases they learn.
- Develop a print-rich classroom that suits their developmental level. Bumper stickers, advertisements, cartoons, posters, and recipes are some examples (Schiffini, 1996).
- Use the Language Experience Approach (see Chapter Five) and other ways to do shared reading and writing.
- Teach to your students' interests.
- Find ways to let students share their developing English skills, such as reading simple books to kindergarten or first-grade students.

Motivation

Motivation to learn English that is internally driven is more powerful than externally driven motivation. Internal motivations can include a desire to make or play with friends, to understand popular music, or to embrace the culture. External motivations may include a parent's desire for the student to learn English, or the desire to learn just enough English to get by with the hope of returning to the home country (Lightbown and Spada, 1999).

Aptitude

You may find that some of your students seem to quickly pass through stages of English development while other students proceed more slowly. Some individuals have an innate aptitude for language learning that can account for different rates of development among your students.

We have discussed factors that impact your English learners outside of school. Now let's look at school-related factors that influence your students.

The School Environment

Your school should have a welcoming tone and atmosphere for all visitors. For parents and families of English learners, there should be materials available in other languages. There should be support for classroom teachers such as up-to-date English as a Second Language curriculum materials and opportunities for professional development in this area. If the school has a large English learner population, the school should have an appropriately credentialed teacher to work with English learners as well as to serve as a resource for other classroom teachers. There may even be a newcomer's class or, in a large district, a newcomer school exclusively for newly arrived non-English speaking immigrant students. In this environment, students learn basic English and receive an orientation to the culture of the United States. Ideally, there should be a community liaison at the school to handle outreach to parents. These are indicators that the school takes success for English learners seriously and is committed to actively supporting their education.

Teacher as the Creator of a Positive Classroom Environment

A positive teacher-student relationship is the fundamental and necessary building block for learning. Students will want to learn when they respect and trust you, and when they believe you have their best interests in mind. A welcoming environment will also help students feel a sense of security and belonging.

Respect for the primary language and culture

Your goal is to help your students learn English and grade-appropriate subject matter. At the same time, it is important to show your students that you respect their primary language and to model for them that you, too want to learn phrases in a new language. For example:

- Learn to say their names correctly.
- Avoid the temptation of giving your students English language names unless they make this request.
- Invite your students to share basic phrases in their language—*good morning, good-bye, see you tomorrow,* and other phrases that everyone can learn and enjoy using on a daily basis.
- In addition to the usual—posters, maps marking students' origins, and playing international music—label classroom objects in the languages your students speak and in English.
- An imbalance in numbers of students from a particular language background might make it easy to give a single student's language a quick pass-over. Be certain to include everyone.

Simply stating the main idea in this paragraph cannot convey the importance and weight of incorporating it in teaching English learners: *Make diversity and learning about diversity part of your curriculum.* Make multicultural education the standard—visually, aurally, and within the curriculum.

- If you do not have time to study a particular culture, country, or region, you and your students can do daily or weekly "fast facts"—a few quick facts about aspects of a student's culture.
- Let students share special holidays and events.
- Encourage them to bring items from home that help other students understand aspects of that student's culture.
- Encourage students to write bilingual stories either at school or at home with their parents' assistance and allow them to share these at school.

Literally hundreds of research studies show that learning climates that are respectful and inclusive of students' linguistic and cultural backgrounds help them to succeed academically. Conversely, years after schooling experiences, students remember when they have been belittled, alienated, made to feel ashamed, or embarrassed about being different, as well as frustrated at not being connected to the curriculum. As part of their research study, Dunlap and Weisman (2005) asked teachers to name the challenges that non-white students must overcome to be successful in school. Consider these two sample responses from teachers:

> "Non-white students have difficulties with not just the language but traditions . . ."
>
> "Yes—language is a very obvious one, but they also have to overcome the challenge of culture. They celebrate days that we don't, for example, Thanksgiving. We assume our children are familiar with these events, but they are not."

Inviting and encouraging your English learners to have a voice in your classroom and including their language and culture will assist them in becoming successful students and endear you in their memories for years to come.

Buddies

A friendly and inviting way to orient a new English learner is to buddy him or her up with an outgoing and nurturing English speaker. Establish a "welcoming committee" of English-speaking students who you select and groom for this assignment. Arm them with school maps and key phrases, such as: "This is the bathroom." "Here's the lunchroom." Have them show the English learners classroom routines like how you have your students line up, where the writing paper is kept, and so on. The members of your welcoming committee will enjoy their important job, while the English learners will be able to establish a friendship as well as get anchored in the life of the classroom and school and learn some English along the way.

Language embedded in routines

Use daily and weekly routines as scaffolds (Peregoy and Boyle, 2005). By using these routine classroom procedures, you will help English learners understand how the day and week flow, as well as learn vocabulary associated with these routines. Every day the student hears phrases like "OK, let's start picking up. It's almost time for the first row to get their going home papers from their cubbies," or "Will the two line leaders please pass out the lunch tickets?" These phrases are embedded in specific contexts at certain times of the day. Your English learners will learn these phrases and their meanings, adding these to their expanding vocabulary and understanding. A visual calendar would be a valuable tool in the beginning weeks of the school year, as students learn classroom routines. Create weekly or daily calendars that match clock times with subject areas and tape it to the students' desks. Make them as simple or as detailed as you desire. An example of a weekly calendar is provided in Table 2.1 on page 37.

Table 2.1: Sample weekly calendar

	Monday	Tuesday	Wednesday	Thursday	Friday
8:00	Roll Lunch Count Pledge	Roll Lunch Count Pledge	Roll Lunch Count Pledge	Roll Lunch Count Pledge	Roll Lunch Count Pledge
8:20	Language Arts Groups	Language Arts Groups	Language Arts Groups	Language Arts Groups	Language Arts Groups
10:00	Recess	Recess	Recess	Recess	Recess
10:30	Writing Workshop	Library Time	Writing Workshop	Computer Lab	Writing Workshop
11:00	Math	Math	Math	Math	Math
12:00	Lunch	Lunch	Lunch	Lunch	Lunch
12:40	Silent Sustained Reading	Silent Sustained Reading	Silent Sustained Reading	Silent Sustained Reading	Silent Sustained Reading
1:00	ELD	Science	ELD	Science	ELD
2:00	Social Studies	PE	Social Studies	PE	Art

Parents as resources

Find ways to invite and use parents as resources in the classroom. Ask them to share cultural items, assist with fieldtrips, or supervise small groups of students. A teacher we know often had parents just drop by the classroom. Sometimes she had tasks for them and sometimes she didn't. In order to make use of their time, she would pull out her enormous basket of magazines and ask the parents to look for and cut out categories of pictures—pictures of food, pictures of people engaged in activities, pictures of houses, pictures of red objects or blue objects, etc. The teacher was either able to use the pictures for her picture file or use them with student projects. Parents can also provide enrichment by explaining holidays and demonstrating traditions native to their culture.

Let's now turn our attention to matters specifically related to English language development.

Two types of language use

In Chapter One we stated that language is used for different purposes. We said that classroom language can be used for social purposes or academic purposes. James Cummins (1981) has provided a very useful way of capturing this difference. He refers to social language use as "basic interpersonal communication skills"—BICS—and academic and cognitively demanding uses of language to teach and learn concepts as "cognitive and academic language proficiency"—CALP. These are important distinctions because they help teachers understand that levels of English learner proficiency are directly related to how easy or difficult uses of these types of language will be for their students.

Here and now examples of social language, whose meaning is context-linked and obvious, include playground and lunchtime language. We say *here and now* because the language used is obvious within the context at that moment in time. Phrases such as "Don't push me" or "Do you want to look at this truck I brought to school today?" are all examples of social language that students hear many times and whose meanings become obvious given the context in which they occur. This type of language is vastly different from the abstract academic language used in instruction that typically has no context clues to support meaning. For example, an English learner has no context for understanding the notion of "taxation without representation" or a lecture about the historical build-up to the French Revolution. The difference in these uses of language also explains why a student may be able to get along just fine in English at recess or at lunchtime but has difficulty understanding content-area instruction.

We are going to ask you to do an exercise here that will help you internalize the difference in these uses of language, again borrowing from the work of Cummins. In Figure 2.2 you see what is referred to as *Cummins' Quadrants*. There are two continua—a vertical one running from top to bottom indicating *cognitively undemanding* to *cognitively demanding* language. The horizontal continuum runs from left to right, intersecting the vertical continuum, and it represents context-embedded to context-reduced language. *Context-embedded* language is *here and now* language, like we mentioned previously. That is, it is language that is understandable because the context makes the meaning obvious. *Context-reduced* language is abstract and has nothing in the environment to refer to that helps make the language understandable.

Figure 2.2: Cummins' Quadrants

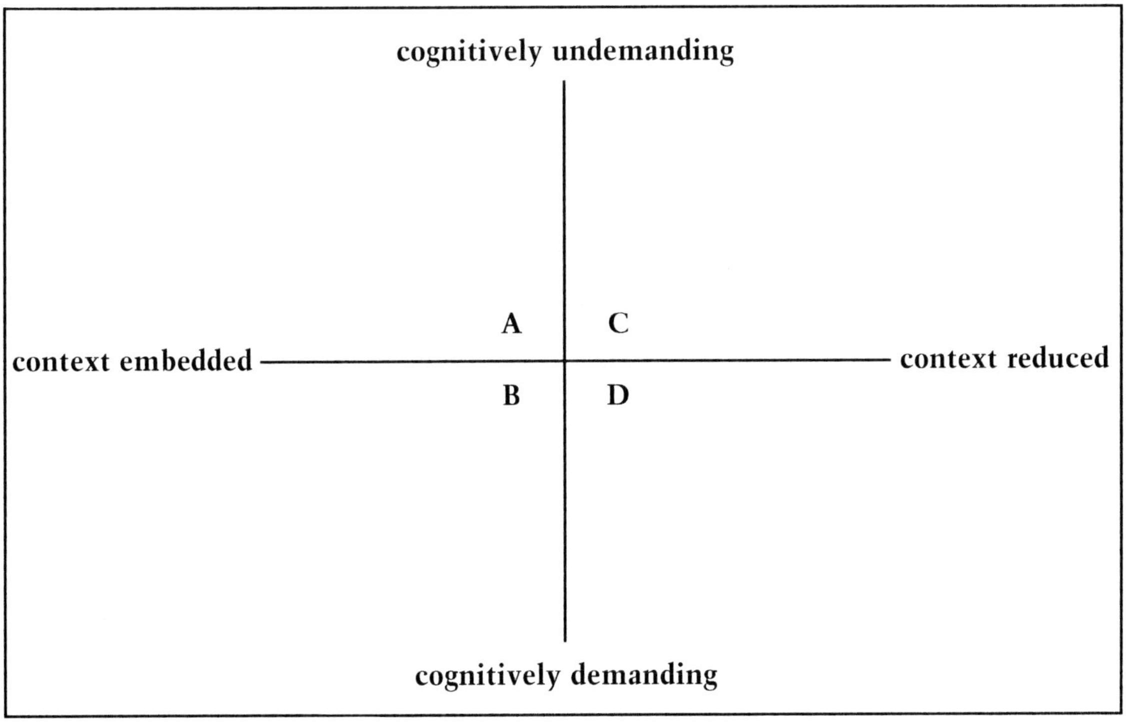

Now to the activity. Below is a list of eight activities and subjects that can occur in a typical school day, let's say, in a fifth grade class.

1. **Spelling**— *-ly* and *-ior* endings
2. **Recess**
3. **Social Studies**—life in the early colonies
4. **Math**—adding fractions with uncommon denominators
5. **Lunch count** and **Pledge**—Opening activities
6. **Science**—respiration in living things
7. **Literature**—*The Witch of Blackbird Pond*—silent reading of chapters four and five
8. **Word search**—vocabulary taken from *The Witch of Blackbird Pond*

In Figure 2.2 above you see the quadrants where you'll place each of the activities and subjects—in A, B, C, or D. Place each bold-faced activity or subject from the list above in the quadrant you think it belongs in, depending on how *context embedded* or *reduced* and how *cognitively undemanding* or *demanding* it is. Remember that each line represents a continuum of context and cognitive engagement.

Here's how we placed the activities in the quadrants:

Figure 2.2: Cummins' Quadrants

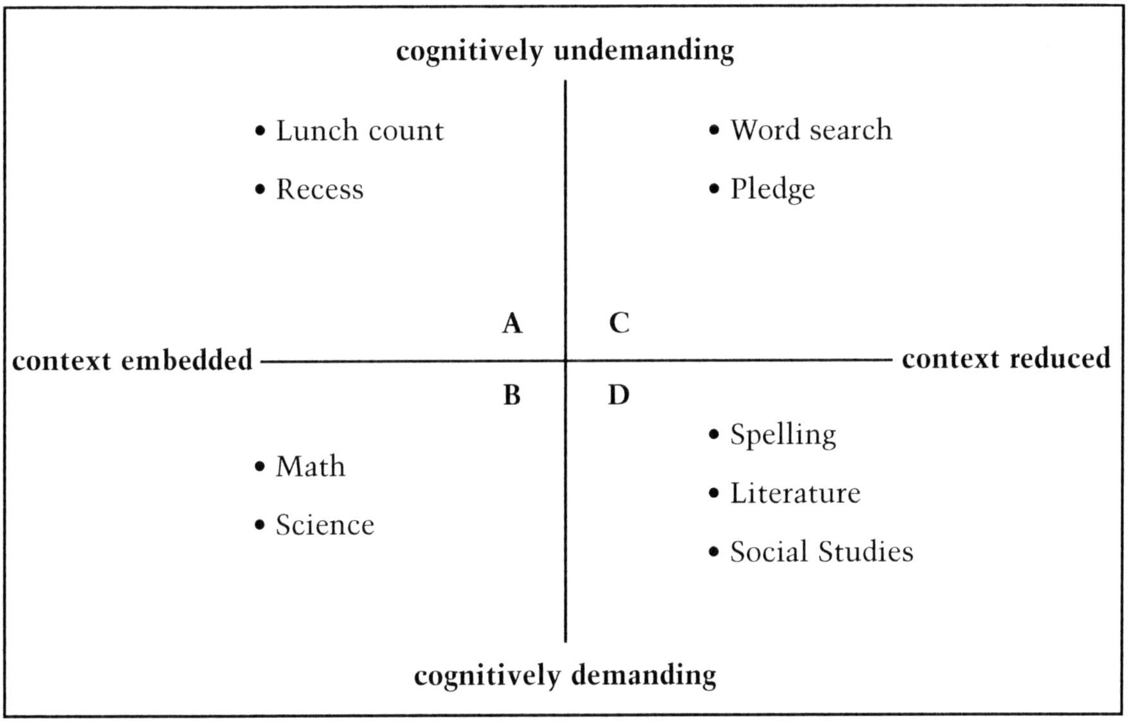

Now to explain why we placed each activity or subject where we did. First, let's assume that our English learners are at an intermediate stage of development. Their social and oral language development is quite good, but they lack a great deal of academic language that will allow them to perform well with cognitively demanding tasks that do not provide contexts or scaffolds to help them understand.

1. **Spelling.** We placed spelling toward the top of quadrant D because understanding the spelling patterns of how nouns can change to adjectives by adding *-ly* and verbs to nouns by adding *-ior* is somewhat cognitively demanding. There is no

context to help students internalize these patterns, unless a teacher or a fellow student takes the time to point them out and helps students understand what the words mean and how words in English can change by adding these endings. This could be a helpful activity for students, depending on the support received. In fact, it could be a very good ESL/ELD lesson.

2. **Recess.** Recess belongs in quadrant A. It is full of social language that is cognitively undemanding and has a great deal of *here and now* context-embedded language like "Throw me the ball" and "You go first." Recess, lunchtime, and transition times ("Please put away your spelling books." "I want to see how quickly you can open your social studies books to page 98.") provide good opportunities to develop social language because they are rich in context clues that help an English learner hear phrases repeatedly and thus learn them easily.

3. **Social Studies.** We placed social studies in quadrant D because social studies texts are typically dense with concepts and new vocabulary. Unless the teacher works at finding ways to make these abstract concepts obvious, the English learner will be challenged to understand.

4. **Math.** We placed math in quadrant B. If the teacher uses lots of visuals like fraction pieces that the students can manipulate, does many sample problems on the board, talks while explaining and showing, and elaborates on explanations, then the learning is context-embedded. The learning and language are cognitively demanding and the teacher provides comprehensible language by making the lesson context-embedded with material and visuals that support the language of the lesson.

5. **Lunch count and Pledge.** The two opening activities are divided between quadrant A and quadrant C. The daily lunch count has lots of context to make the meaning comprehensible but is cognitively undemanding so it belongs in quadrant A. The pledge of allegiance is placed in quadrant C because it has no context other than standing in front of the flag with hand over heart, and for an English learner it is a string of incomprehensible words. For many students, the pledge is cognitively undemanding because they are typically not expected to internalize its meaning.

6. **Science.** We placed the study of science (respiration) in quadrant B. While there are new vocabulary words and new material to read, if the teacher uses graphics and charts, the concepts are context embedded. The students will do an experiment that will challenge them to think about ways to improve respiration in the body.

7. **Literature.** Literature belongs in quadrant D. Just like social studies, reading for understanding in a language while trying to learn it is very demanding and often frustrating. In this case, the students are asked to do silent reading. Unless the teacher has taken time and effort to provide comprehensible input via focused vocabulary teaching and a visual preview, such as a story map of what happens in chapters 4 and 5, reading a piece of grade-appropriate literature will be challenging to an intermediate and even an advanced English learner. This type of reading requires a teacher to use many instructional "tools" that we will discuss in the following chapters. Literature and content-area texts, especially social studies, are examples of cognitively and academically demanding tasks

that will frustrate the English learner unless you take the time and effort to help him or her understand.

8. **Word search.** We placed the word search in quadrant C. There is no context to understand what it is supposed to mean, so it is context-reduced. Furthermore, finding words in a random display of letters is not cognitively demanding.

We hope that working with Cummins' Quadrants has given you a sense of the different uses of classroom language, what is comprehensible to an English learner, and what activities need support to be understood. We also hope you understand that different types of language serve different purposes, and that a student with a fairly good grasp of social uses of language (BICS) will probably not have an equally developed use of academic language (CALP). Finally, notice that reading literature and heavy-laden conceptual text in English will require you to scaffold meaning and understanding for your English learners.

Aspects to Consider in Your Classroom

As you consider providing meaningful social and cognitive support, you will need to think about the following:

- the number of English learners you have in your classroom
- the stage of English language development for each student
- the number of non-English languages represented in your classroom
- the type of primary language support you or others can provide your English learners

These are important because they affect ways in which

you can organize your students to provide support for concept and language learning.

Number of English learners you have in your classroom

This matters because it affects how many native English speakers can provide linguistic models that can work with English learners. If you are the only native English speaker, the demand on you will be high. Conversely, if there is only one English learner, the temptation will be to skip over this student's English learning needs and focus on instruction for your native English speakers. For this reason, we promote the concept of a "critical mass" of English learners—say from four to ten—placed in one classroom with a teacher who holds the proper professional preparation and dispositions to work with English learners.

Stages of English language development

This matters because it will affect how much time you need to spend preparing comprehensible content instruction. If your students are all at the same level of English development, let's say intermediate, then you will prepare lessons with the kind of support to meet their learning needs. If you have beginning, intermediate, and advanced levels of English learners, each level will require somewhat different approaches, or tools, to help them understand content lessons. In Chapter Seven, we provide a sample fourth grade social studies lesson that plans instruction for all levels of language development. It serves as a model for you to use as you plan content instruction for your own classroom.

Number of non-English languages your students speak

Having two or more speakers of the same non-English language can help you and your students, especially if one is a more advanced English learner. It is a definite advantage to allow them to use their native language to help one another understand and negotiate content meaning. After all, your goal is to teach, and so much the better if you have another person who can help you convey the content in a way that your English learner understands! Your students' English language development and learning will be greatly enhanced when you provide opportunities for students to help one another. In Chapter Seven, we will give you some specific tools to help you accomplish collaborative student work. We offer a caution, however—*do not permit constant translation.* First, it takes away learning time from the bilingual student who can become understandably resentful. Second, it causes the student with lesser-developed English skills to become dependent on someone for translation and less willing to work at negotiating for meaning in English.

Type of primary language support you or others can provide your English learners

You may find yourself with the advantage of assistance from a bilingual aide or parent, or you may have some working knowledge of your students' primary language. Take full advantage of all the primary language support you can garner for your students. In the paragraph above, we mentioned student-to-student language support. Any type of primary language vocabulary or concept preview or review you can provide for your English learners will help them a great deal. The role of the primary language is to support learning that serves to strengthen a student's education. Whatever the student understands and learns in the primary language can be expressed in English once he/she learns that vocabulary in English. This is the concept that undergirds bilingual education. We mention again that constant translation is *not* bilingual education, nor is it sound educational practice. When implemented correctly, bilingual education is a powerful way for a student to become bilingual and biliterate. If a good bilingual education program is not available at your school, you should use all primary language resources available to you for the benefit of advancing your students' learning.

Ways you can organize your students to provide support for concept and language learning

We stated in Chapter One that language is learned in social contexts. Aspects of social learning include an interesting environment that invites students to learn, a more knowing guide who knows how to listen and explain, and an actively participating learner. Learning buddies and cooperative learning groups provide social structures that are highly beneficial for English learners. A great deal of educational literature is available on ways to organize for cooperative learning. We will discuss some of these in Chapter Seven. Cooperative learning strategies allow for clarification within small groups of students, an opportunity to repeat and practice language, to hear content language modeled and expanded by more fluent peers, and they require participation of everyone, including the English learner. We encourage you to use cooperative learning structures in your classroom if you do not already. They are of great benefit to all your students.

Having considered the many influences that affect your English learners, we arrive at the question you've probably had in mind all along.

How Much Time to Acquire English?

If we first consider a native English-speaking student, we might agree that by grade six or seven, the student is able to listen, speak, read, and write at a fairly good level of proficiency. A student is able to accomplish this proficiency in seven or eight years of formal education—from kindergarten to sixth or seventh grade. Education beyond sixth or seventh grade, of course, continues to refine the student's abilities in the four domains of language. Research shows that about the same amount of time is necessary for an English learner to acquire fluent English language proficiency in the four domains—between five to eight years (Collier, 1987, 1989; Ramirez, 1992). Just as with native English speakers, less time is required to develop speaking and listening skills. However, fluent literacy skills—reading and writing—simply take more time. We remind you of the last point we stated in Chapter One—language learning takes time!

Important Points to Remember

- You, the teacher, are one of many sources of influence.
- Cultural and family expectations play a significant role in student learning.
- Seek primary language support for your English learners at home and at school.
- Student learning is positively influenced by classroom respect for home language and culture.
- Help students develop social and academic uses of language.

Now you're armed with knowledge of the many influences that affect your English language learners. In the next chapter we'll look at ways you can assess your English learner, and in the following chapters we'll provide many specific teaching tools for you to use that will help your student along the path to English language fluency and subject matter understanding.

Post-Reading Reflection

1. What factors besides learning English are important influences on your English learners' ability to learn?

2. How would you respond to a colleague who maintains that limited English-speaking parents should speak English to their children at home?

3. What can you do in your classroom to promote respect for all students' culture? Why is it important to do so?

Chapter Three

Assessment

> *"Would you please tell me which way I ought to go from here?" asked Alice.*
>
> *"That depends on where you want to get to," said the cat.*
>
> *"I don't much care where," said Alice.*
>
> *"Then it doesn't matter which way you go," said the cat.* (Lewis Carroll)

This amusing quote from *Alice in Wonderland* captures the importance of assessment. We want to make an instructional plan for our students, provide instruction that helps them achieve, and then find out if they've gotten to the point we've planned for them.

Many states have developed their own formal assessment tool for English learners. This type of standardized test is typically given on a yearly basis by someone who has received special training on administering it. These

English language assessments are important because they do the following:

- screen and identify students who need English language instruction
- establish appropriate placement for level of instruction
- reclassify students to move to a different level or exit the English learner program
- monitor English language development
- provide information on program evaluation
- establish instructional and student accountability (O'Malley and Valdez Pierce, 1996)

This information is certainly useful for a district to maintain and may be required by state law. The results of these standardized tests provide an *indirect measure* of your students' English language ability because it is removed from actual tasks a student does in the classroom.

Classroom teachers don't usually find these indirect measures so useful for instructional purposes in the classroom. First, you may receive this information briefly reported by a phrase or a number such as "early advanced fluency" or "Level 2" depending, of course, on the test that's used and its rating scale. This may not help you understand exactly what level of academic performance you should expect from your student. Second, there may be a lapse of time from the point that the test was administered to the time you receive the results. The results of the test are a snapshot in time reported to you after the student has already made some progress in your classroom. The results may not adequately capture your student's current level of English language development.

Classroom-Based Assessments

What can help you a great deal as a teacher of English learners are *direct measures* that you can obtain yourself from your own classroom-based assessments. These help you answer two important questions: (1) *What level of performance on academic tasks should I expect of my English learners?* and (2) *What support do my students need to help them learn and perform well on academic tasks?* Furthermore, informal and direct measures can help you answer another question: *How is my student's English language development progressing?*

Diagnostic assessment, as the term indicates, is a diagnosis of areas of need. It helps you select the types of support your students will specifically need with listening, speaking, reading, and writing to help them achieve both academically and to improve their English language abilities. This support will also help them improve future performance. Teachers of English learners find direct measures the most helpful. These include classroom-based measures of student performance, such as writing samples, projects, student-made exhibits, and other activities that students engage in as a part of regular classroom activities. Assessing this kind of student performance gives teachers direct classroom-based feedback.

In this chapter we're going to provide you with some classroom-based assessments that you can use to give yourself direct measures. You can use these for diagnostic purposes, for academic assessment, and for helping to track your students' language development throughout the school year.

Oral Language Assessments

1. SOLOM: Student Oral Language Observation Matrix

This test was developed some years ago by the Foreign Service Institute and adapted for use with school-age students by the California State Department of Education. It is easy and useful for you to use with your English learners and it will give you a quick snapshot of your English learner's oral proficiency. This matrix can be found online at:

http://www.helpforschools.com/ELLKBase/forms/SOLOM.shtml.

Second language learners are usually not equally proficient in all aspects of language. This quick oral language test, which is found on the web site above, will help you understand your students' strengths and areas of greatest need. Across the top of this matrix are numbers from one to five, indicating levels of proficiency, with five being the most developed. Vertically and along the left of the matrix are dimensions of language, including C*omprehension*, *Fluency*, *Vocabulary*, *Pronunciation*, and *Grammar*. Each coordinate in the matrix contains a descriptor that corresponds to the level of proficiency in that dimension of language. The teacher listens to the student's language, then marks an "X" in the cell that best describes the student's abilities for each of the five dimensions of language. For example, you may rate your student as a level 2 in comprehension, 2 in fluency, 3 in vocabulary, 3 on pronunciation, and 2 on grammar, for a total of 12 points out of a possible 25.

It is important to do this assessment as soon as possible in the school year to establish a verbal baseline for your student. And, of course, make the interview as relaxed and casual as possible. We suggest you keep an audiotape of each of your English learners. If you can find time to tape record your student two times a year, you will be able to hear a substantial growth in verbal development, and your students will delight in hearing how they sounded some months earlier.

We also suggest that you consistently use the same questions for this oral assessment. These provide a stable baseline for judging language production across students and across time. You will be able to hear, for example, how a student is able to offer an expanded response to a question to which, months earlier, he/she could provide only a limited response.

Here are questions that we've developed for you.

Sample Questions

1. What's your name?
2. How are you today?
3. Can you tell me what day it is?
4. Show me the . . .
5. Who is your best friend?
6. Tell me as much as you can about your best friend.
7. What do you like about school?
8. What have you been studying about in school?
9. Tell me about some things you really like to do.
10. Talk about what you like to do best at recess.

The first five questions will allow even a very limited speaker to respond and avoid embarrassment or frustration. The latter five questions permit a student to elaborate and to demonstrate verbal ability and growth over time. Notice we stayed with topics that are school related. This is because a student may not have the English vocabulary to discuss topics from home. In addition, some questions may not be appropriate in the home culture. An example is "How do you celebrate your birthday?" because not every culture follows the custom of celebrating birthdays.

2. Social and Cognitive Listening and Speaking Anecdotal Record

A second assessment of oral language ability is this observational checklist, based on your active listening skills of one student for a day or two. It is developed with the social and cognitive uses of language in mind and assumes that social uses of language will precede cognitive language uses. It also helps you establish how comfortable your English learners are becoming as they use their new language with increasing numbers of people. The tendency of second language learners is to be somewhat timid in the early stages of second language development and then more willing to speak in larger groups as their ability and ease increase. Completing this observational checklist twice a year will provide evidence of oral English language growth. A blank copy is provided on page 57.

Social and Cognitive Listening and Speaking Anecdotal Record

Directions: Observe your student for one day. Then complete this form by circling the appropriate categories.

Student name: ______________________________ **Date:** __________

COGNITIVE DOMAIN (vertical axis) · SOCIAL DOMAIN (one-on-one → whole group)

Skills being observed	Comments
1. Responds to instructions one-on-one small group whole group	
2. Gives instructions one-on-one small group whole group	
3. Talks about/responds to a story one-on-one small group whole group	
4. Describes real events one-on-one small group whole group	
5. Describes imaginary events one-on-one small group whole group	
6. Answers a question (lower level) one-on-one small group whole group	
7. Asks a question (lower level) one-on-one small group whole group	
8. Answers a question (higher level) one-on-one small group whole group	
9. Asks a question (higher level) one-on-one small group whole group	

Printed by permission from Carmen Zuñiga Dunlap.

Other Assessments

The Important Thing

Our colleague Dr. Teresa Crawford has developed a pre- and post- assessment that is highly adaptable for all areas and all levels. She bases it on *The Important Book* by Margaret Wise Brown (Harper Trophy, 1990). The book discusses many common objects—a spoon, a daisy, grass, and so on. The first sentence about each object is "The important thing about . . . is that it is . . . ." ("The important thing about a daisy is that it is white.") This statement is followed by three or four statements about its other attributes—"It is . . . It is . . . It is . . ." The last statement is a repeat of the first. "But the important thing is that it is . . . ."

This language pattern can be understood even by beginning level English learners. It can be used with small groups, large groups, or individuals. It is highly adaptable in that it lends itself to any topic or subject matter. Individuals can create their own important attributes about an object, which will likely differ from someone else's. The assessment lies in what the student selects to talk about and describe. It is an excellent form of pre-assessment of your English learners' knowledge and vocabulary before beginning a unit of study as well as a post-assessment indicating what they have learned.

Retellings

A retelling is a written or spoken recall of what a student remembers after reading or hearing a piece of text. It is a type of *performance assessment* which requires the student to construct a response, engage in higher order thinking, and use the four domains of language—listening, speaking, reading, and writing. A retelling is also considered an *authentic assessment* because it parallels a real-life activity that someone may be expected to do outside of school. (O'Malley and Valdez Pierce, 1996).

Retellings are useful tools for teachers and for students. They can indicate to a teacher what a student:

- remembers;
- thinks is important to remember and include;
- can infer from the text; and
- connects from the text to his/her own experiences.

A retelling also indicates how a student:

- uses language;
- constructs meaning;
- sequences and organizes information; and
- processes and connects large pieces of text.

A retelling can indicate a student's ease of expression either orally or in writing. It can also document a student's present level of competence and, with repeated opportunities to do retellings, document improvement over time. A retelling is excellent instructional tool. The use of retellings over time helps students develop comprehension and improve their concept of a story (setting, plot, and resolution), vocabulary, ease of oral expression, and correct use of written language.

A retelling is a quick and highly flexible assessment tool. You can ask a student to do a retelling of a story or expository text taken from a textbook. You can ask the student to do this in English or in his/her primary language. A retelling done in the primary language can give you a general idea of the student's comprehension and oral or written fluency. If the student reads then writes the retelling, this can be useful in determining a student's literacy level in the primary language, indicating what resources the student can draw on in becoming literate in English. The student can read the text or listen to someone else reading and then do the retelling orally or in written form. Finally, a retelling can be done with a teacher, classroom assistant, a fellow student, or a cross-age helper. Given the ease of doing retelling and all the benefits of having your students do repeated retellings, you can see the advantage of using them for assessment purposes as well as an ongoing part of your instruction.

We've developed a retelling rubric that you can use just as you see it here or that you can modify to suit your own purposes. This rubric is called the Retelling Profile and it is provided on page 61.

Retelling Profile

Student ______________________________ Date____________

Story title and author ______________________________

or

Subject, text, and page numbers ______________________________

Check one:

_____ text was read aloud to student
_____ text was read silently by student
_____ text was read aloud by student
_____ text was read aloud with a student partner
_____ other:

Scoring: 0 indicates no evidence
1 indicates limited evidence
2 indicates moderate evidence
3 indicates much evidence

	0	1	2	3
Text-based evidence				
Recall of important information				
Recall of main ideas				
Recall of supporting ideas				
Recall of sequence				
Language use Retelling demonstrates student's use of. . .				
Vocabulary				
Sentence structure				
Language conventions and mechanics				
Standard spelling				
Reaction and response to text				
Personal observations				
Creative impressions				

Printed by permission from Carmen Zuñiga Dunlap.

Two teachers we know modified this rubric to include elements that appeared on their district's new report cards for which they had not previously gathered data in their teaching. Including these elements as part of the retelling rubric and engaging their students in repeated retellings gave them data by which to judge student progress and, as a result, information that they needed to assign grades.

These teachers, Jamie and Laurie, were students in our program. They did a classroom-based retelling project for a graduate course that we taught. For their project they devised a creative combination of buddy journals (Bromley, 1995a) and retellings. They each taught third grade in the same school. They buddied their students across the two classes—a native English speaker with an English learner. Each student maintained a journal. Twice a week, the students would write to each other. The first time was on a topic of their choosing and the second time was a retelling. Twice a week, after each entry, the teachers bundled up the journals, exchanged them, and distributed them to the buddies in the other classroom. The buddy would write back and the teachers would exchange the journals, returning them to the owner who would read and respond to what the buddy had written.

Weekly, the teachers agreed which story from the Scholastic Literacy Program they would read to their students. Each teacher would then read the same story to her class and have the students do a written retelling of the story. The teachers would then exchange the journals, and each student would respond to the buddy's retelling. An initial assessment of their students' retellings revealed useful information on their English learners' strengths and weaknesses. Jamie and Laurie tracked their students' progress for "five short weeks" as they reported, assessing the retellings based on our Retelling Rubric. In just five weeks, they stated that "our students grew in their reading and writing [abilities]. The retelling journal was an excellent resource for many things . . . valuable information about our students, about ourselves in our teaching, and timely assistance with our report cards. We would highly recommend this tool in any teacher's classroom."

We hope this idea from Jamie and Laurie has given you encouragement to use retellings and to adapt them to meet your particular needs.

Final Suggestions for Assessing Your English Learners

Now that we've given you some assessment tools you can use with your English learners, we want to add a few additional thoughts and cautions about classroom assessments that you create for them. Try to create content-area assessments for your English learners that do not require them to rely so heavily on reading and writing to demonstrate what they have learned. This is especially important for beginning- and intermediate-level students. Can they create a time line? Create graphic organizers and add phrases and important vocabulary to demonstrate their knowledge? Demonstrate their knowledge by creating an experiment? These are just a few ideas as you expand your thinking on other ways to approach assessment for your English learners.

Important Points to Remember

- Use classroom-based assessments that give you diagnostic information about your students' English language skills.
- Assess your students' English language skills just as soon as possible in the school year to determine a baseline and at the end of the year to establish growth in their abilities.
- Retellings are easy and adaptable assessment tools that provide diagnostic information, and when used repeatedly, tools that help students develop many literacy skills.
- Find ways to assess your students' content knowledge that do not depend greatly on reading and writing, especially for beginning- and intermediate-level English learners.

Post-Reading Reflection

1. Describe how you will use one of the classroom-based assessments in your classroom.

2. What is one way you could buddy up with another teacher to help your English language learners?

3. What's important to remember about assessing English learners' content knowledge?

Chapter Four

Oral Language Development

Consider the following scenario from a fifth-grade classroom:

Diana's students were engaged in a Know-Want-Learn activity on the Westward Movement. This strategy engages students in articulating what they know about a topic, what they want to know, and a post-study discussion on what they have learned. Various students contribute to the "know" and "want to know" sections of the large chart taped along the length of the blackboard. Diana asks several English learners what they would like to add. Within the context of questions about covered wagons, an English learner raises his hand. Diana calls on him and he simply says "Why. . .?" as he moved his hand in a choppy circle. Diana says, "Oh! Why did the wagons form a circle at night?" The student smiles and nods as Diana writes it on the chart. "Good question," adds Diana (Zuñiga & Yopp, 1996).

Here we see a student who is at an early stage of English development. Diana accepted the student's intended meaning, complimented him for his contribution, and provided a verbal and written model for expressing his intended meaning. Diana is a wise teacher who uses many instructional tools to help her students develop English.

Oral language plays a key role in daily communication. It is also a building block for developing reading and writing skills and for understanding new concepts. In this chapter, we focus on oral language development and provide tools that you can use in your classroom to help support your English learners' acquisition of listening and speaking skills. We begin with a brief discussion of some of the major ideas that have influenced the field of second language teaching.

Theoretical Foundations

For many years, views about teaching a second language centered around behaviorist theories that emphasized the use of pattern drills, memorization of verb tenses, and practice with sentence structures. But this approach was not effective in developing fluency in the second language (Lightbown & Spada, 1999). We now know that what promotes fluency are comprehensible language input and extensive opportunities for interaction and authentic communication.

Steven Krashen's theories have been influential in the field of second language teaching (1982). He suggested the following:

- Teachers should recognize that there might be a silent period when students are not yet ready to speak in the new language; however, they are taking in a great deal of information through listening. Students should not be forced to speak during this silent period.

- The three most important elements related to success in a second language are (1) a low anxiety environment, (2) high motivation, and (3) self-confidence and self-esteem.
- Teachers should focus on activities that use the language for real communication instead of emphasizing grammar rules.
- Second language acquisition occurs primarily by understanding messages, or receiving *comprehensible input*, in the second language. The input can be made comprehensible by utilizing context, visual aids, and gestures to support meaning.

Some of Krashen's suggestions have been challenged. We'll mention two here. The first is that comprehensible input alone is sufficient to promote fluency. Social interaction—conversational give-and-take—is very important. Negotiating for meaning, not just input, is part of the language learning process. (Hatch, 1992; Long, 1983). This supports the work of Vygotsky (1978), who believed that learning is a social process that occurs through interaction between children and others in their environment. Vygotsky maintained that teachers need to provide temporary support or scaffolds—structures that assist in learning and comprehension until students are able to complete learning tasks independently.

The second challenge concerns the role of grammar in language learning. Some research shows that students *can* benefit from grammar instruction within a communicative-based program (Lightbown & Spada, 1999). Including some grammar instruction within meaningful communication activities seems to be most helpful for older learners—ten years old and beyond—who are at an advanced stage of language development.

In Chapters One and Two we discussed the different uses of language, specifically social and academic uses. Cummins' (1994) work suggests that second language learners can develop social language within a few years, but it will take much longer for them to acquire the academic language that is necessary for success in school. It may take as long as five to seven years, or longer, to develop academic language proficiency (Collier, 1987; Cummins, 1994).

First and Second Language Acquisition

First and second language learners progress through very similar stages of language development. However, there are differences. Let's look at some similarities and differences between first and second language acquisition. Babies learning a first language progress from babbling to single words then two and three-word utterances within a span of about two years (Lessow-Hurley, 1999). During this time they are also learning how language works and how to use it as a tool for communication and expressing thought. First language learners gradually develop the ability to express their needs, make requests, share their ideas, talk about the past and future events, and so on. By the time they enter school, children generally have a good understanding of the sound system (phonology), how words are formed (morphology), and word order (syntax) in their native language. Native English speakers and English learners are already aware of how their primary language works and how to use it for communication purposes. The difference is that English learners need to learn how to express what they know in the second language. A second important difference is that school-aged English learners need to acquire academic concepts at the same time that they are developing proficiency in English. That is, they need to *learn* English and learn *in* English. These differences are illustrated in Table 4.1.

Table 4.1: Comparison of First and Second Language Acquisition

Native English Speakers	*English Language Learners*
Understand how English works and how to use it to convey meaning both at home and at school	Understand how language works and how to use language to convey meaning
Enter school with solid foundation in phonology, morphology, and syntax in English	Enter school with solid foundation in phonology, morphology, and syntax in the primary language, but need to develop these in the second language
Need to develop academic concepts in a language they already can understand and speak	Need to develop understanding and speaking skills *as well as* academic concepts in the second language

Labeling Stages of Language Development

As children learn English, they progress through stages of language development. Various programs for teaching English as a second language may have different labels for these stages. We also know that states differ in their descriptions of levels of English language development. For example, California schools use five labels to describe or identify English learners' stages of development: Beginning, Early Intermediate, Intermediate, Early Advanced, and Advanced. The important thing to remember is that, although labels may differ, the developmental process of second language acquisition is the same. Think of language development as a continuum that ranges from the very beginning to a very advanced level of language proficiency. Then, think of points along the continuum that have been given labels. Labeling these points helps us understand what our students' strengths and needs are as they progress along the path of language development. Throughout this book we use three stages of English language development: Beginning, Intermediate, and Advanced. We will describe these stages and provide teaching tools for you to use at each of these levels.

Beginning Stage

The beginning stage is a period when children develop the ability to get meaning from words and phrases and gradually begin to produce the language that they hear around them. Some of the characteristics of the beginning stage include:

- Associating the new speech sounds with meaning.
- Having a *silent period.* It is common for beginners to experience a period during which they do not feel comfortable speaking in the second language. Children may not initially make attempts to communicate verbally; instead, they will indicate their comprehension nonverbally.
- Grasping the idea of a phrase without understanding all of the words—children do this by focusing on key words.
- Relying on contextual clues for understanding key words.
- Gradually beginning to respond using one, then two or three-word phrases.

Students at this stage may experience a silent period, lasting anywhere from a few weeks to a few months. During this silent stage they are developing listening strategies that will form the basis for speaking skills. When students do begin to speak, accept and encourage all efforts. Obviously, your students' early attempts at communication will not be perfect; however, pronunciation and other aspects will improve gradually as they have more practice. We do not recommend exercises to correct pronunciation and grammar at this point.

You can use a variety of teaching strategies with your students at the beginning stage. One of these is Total Physical Response (TPR). Developed by James Asher (1982), TPR is very useful for teaching comprehension at this stage and for teaching new vocabulary at any stage. The basic premise is that language will be acquired more easily if it is associated with body movement. Essentially, the method consists of the teacher giving commands and the students following them. It is important that children have many opportunities to follow the command as they observe the teacher modeling it before they are asked to follow the command without any modeling. Inability to follow the command without modeling means they need more practice listening to it and observing the teacher model it. TPR requires no verbal response, which makes it ideal for beginners, although some children may repeat the commands and may eventually want to lead others in a TPR activity.

TPR can be used to introduce students to basic classroom routines and vocabulary (*stand up, sit down, get in line, open the book, pick up the red paper, pick up the green paper*, etc.). However, you should use it along with other approaches since at this stage students need to be exposed to as much comprehensible input as possible. So, a modified TPR approach should be incorporated into other language activities to teach vocabulary (Krashen & Terrell, 1983). For example, you might have the students draw a picture of their faces and then ask the students to touch various parts of the face. Your teacher talk would include commands as well as a wider range of sentence types. During this teacher talk, *vocabulary needs to be made comprehensible by means of gestures, drawings, pictures, and visual clues.* Your students will not necessarily understand every single word, but they should understand the key words. Here is an example of what an activity might sound like:

> *You've drawn some great pictures of faces. Look at Thuy's face!* (pointing to Thuy's picture). *This is the face and here are the eyes, the nose, the mouth* (pointing to each body part). *You've drawn big, brown eyes. What beautiful eyes! Point to the eyes on your picture. Yes, those are the eyes. These are my eyes* (Teacher points to his/her eyes).

Associating vocabulary words with students' names is another technique that can be combined with TPR to build comprehension and elicit non-verbal responses at the beginning stage (Terrell, 1981). For example, if you wanted to teach the names of fruit, several different types of fruit could be placed in a bag. As each piece of fruit is taken out of the bag, talk about it and incorporate the children's names:

> *Look at this apple. It is a red apple. Can everybody see the apple? It's an apple. I'm going to give the apple to Maria. Maria, here is the apple. Who has the apple?* (Students say "Maria".) *Yes, that's right, Maria has the apple.*

Students can be successful with this type of activity without having to understand every single word. It builds confidence while at the same time develops comprehension.

As students feel more confident and after they have heard English words many times, they will begin to speak quite naturally. When children are ready to speak, continue the same types of activities described earlier but provide opportunities for the children to respond verbally. Terrell (1981) suggests using the following types of questions to encourage beginning students' responses:

1. *Yes/no:* Is this an apple?
2. *Choice:* Is this an orange or an apple?
3. *Completion:* This is a big, red . . . .
4. *Single or two-word questions:* What is this? What are these?

These questions can be woven into your language, which should continue to focus on comprehension, including previously taught vocabulary as well as new vocabulary. Let's suppose, for example, that you have taught color words, clothing words, action words, and animal names but haven't taught words for size relationships. You could use a picture to provide students with an opportunity to say words they have heard many times as well as to build comprehension of new vocabulary. For example:

> *Look at this picture. What do you see?* (Cat) *Yes, here's a cat. And is this a cat or dog?* (Dog) *Yes, it's a dog. Is the dog playing?* (Yes). *Yes, he's playing with the girl. They're playing on the grass. What color is the grass?* (Green). *That's right, the grass is green. What color is the girl's dress?* (Red). *Good, that's right. The dress is red. The dog is big but the cat is little. This tree is big and this tree is little. Is this tree little?*

When students begin to respond verbally, accept their responses, praise them, and expand on their language. If, for example, the student says "play" instead of "playing," you can say," Yes, the dog is playing." In this way, you model the correct response, provide additional language input, and avoid creating anxiety for the student.

Important Points to Remember for Beginning Stage

- Provide comprehensible language input by using gestures, pictures, or real objects.
- Repeat key vocabulary in a systematic way.
- Focus attention on the activity so children become interested and motivated to communicate in English.
- Allow for non-verbal responses until children feel comfortable with speaking.
- When children do begin to respond, accept and praise their responses and expand on them.

Vignette #1

Examine the following classroom vignette. Notice the types of questions the teacher asks of children at different levels of language development. Can you identify four tools the teacher uses to promote language development?

The children participating in this lesson include Miguel and Nuzhat, who have just recently arrived in the United States and do not speak English. The other children have been in school for several months and have begun to produce a few words and short phrases in English. Previous lessons have focused on color words. In this lesson, the teacher is showing pictures of plants and flowers:

Teacher: What do you see in this picture?

Thanh: Flower.

Teacher: Yes, this is a flower. It's a very pretty flower with lots of petals and leaves (points to petals and leaves in the picture). Can you point to the flower? Good, this is the flower. (As she points to the flower). Miguel, point to the flower. Nuzhat, point to the flower. Yes, this is a flower. These are called petals and leaves. And this part is the stem of the flower. What color is the flower? Is it purple?

Children: No!

Teacher: No, it's not purple. What color is it?

(Children hesitate.)

Teacher: Is the flower red or is it yellow?

Children: Yellow!

Teacher: Yes, that's right. The flower is yellow. And the leaves on the flower are green. What color are the leaves, Julie?

Julie: Green.

Teacher: Yes, the leaves are green.

Teacher: (as she points to the leaves) Miguel, point to the leaves. Nuzhat, point to the leaves. These are the leaves.

Tools

What tools does the teacher use?

1. uses pictures to convey meaning
2. repeats key words
3. asks a question that involves choice to encourage response
4. allows for non-verbal responses from beginning-level students

Intermediate Stage

Your students will progress rather quickly into the intermediate stage when their understanding is greatly expanded and they are much less hesitant to speak. Characteristics of this stage include:

- ability to speak in simple sentences and engage in conversation;
- ability to reproduce familiar phrases that are memorized through repeated use;
- ability to relate details of an event or story, identify main ideas, and summarize a plot;
- ability to respond to open-ended questions that relate to *here and now* situations; and
- frequent errors in grammar.

During this stage, continue to expand students' vocabulary and use many of the same techniques described earlier. You can use Total Physical Response to teach more complex language and vocabulary by giving commands such as: *Give the blue pencil to Lily. Point to the crayon that is not yellow. Fold the top left corner of your paper.*

It is still important to repeat key vocabulary frequently and to use visual aids and concrete material to convey meaning. Additionally, encourage students to use more language by asking open-ended questions such as: "Why did you like the story?" or "How did you make that?" Now is also the time to involve students in more linguistically demanding tasks that require extensive understanding and production. For example, model language that describes and compares, then engage students in doing the same. It is also important to continue to provide interesting activities that will motivate students to want to communicate in English. Students can participate in retelling stories and take part in Reader's Theater presentations and games to practice their language skills. A few games are described at the end of this chapter.

At this stage, vocabulary can also be taught through academic content using some of the methods already described. For example, TPR can be used in a math lesson ("Point to the pyramid." "Point to the cube.") or in a science lesson ("Point to the roots." "Point to the stem.") It is important to remember that although intermediate students are able to understand a great deal more than beginners, they will need to hear new words many, many times before they are internalized and can be produced automatically. Therefore, use TPR and other scaffolding strategies described earlier whenever new vocabulary is introduced.

While students in this stage will make many errors in their speech, it is not advisable to constantly correct them. Instead, model correct language form and encourage them to express themselves. When you model corrections you will be providing more comprehensible input that students need to refine their language. At this point, it is neither appropriate nor productive to correct students' grammar and teach grammatical rules.

Important Points to Remember for Intermediate Stage

- Continue to provide comprehensible language input by using gestures, pictures, or real objects.
- Continue to develop vocabulary.
- Focus attention on the activity so that your students become interested and motivated to communicate in English.
- Encourage describing and comparing.
- Ask open-ended questions to encourage more language production.
- Accept and praise student responses and expand on them.

Vignette #2

Read the classroom vignette. Can you identify five tools the teacher uses that are appropriate for these intermediate-level students?

> Susana's fourth graders are lining up for lunch. She turns to the student line and sees Ramón punch Suwat in the arm. "This is not what I expect from you, Ramón. What happened?" Ramón gestures toward Suwat and says, "He stepped my foot." Suwat quickly says, "He cut the line." Susana takes the boys aside. "O.K. Ramón cut in line, so Suwat stepped on Ramón's foot, and Ramón hit Suwat. Is that what happened?" Both boys nod in agreement. "But teacher, you say my name to get in the line and Suwat no," adds Ramón. "Yes, Ramón, I called your name to get in line, and I called Suwat's name to get in line, too. The way you acted toward each other makes me feel sad. I want us all to respect each other. I want us all to be nice to each other. Do you remember the rules we made that we all agreed to follow the first week of school?" Susana points toward the posted chart. The boys look at it and nod in agreement. "Ramón, what should you have done instead of hitting Suwat?" Eyes cast down, Ramón says, "I should get in the line behind Suwat and no hit him." Susana nods and says, "Exactly—get in line behind Suwat and not hit him. And Suwat—how about your behavior?" Suwat, also with eyes cast down, says "I not step him." Susana nods again and says "Yes, that's right. You should not step on his foot. And if this happens again, what will you do?" "We be respect each other?" asks Suwat. "Thank you," says Susana. "You will respect each other."

Tools

What tools does the teacher use?

1. accepts her students' language and models standard language rather than overtly correcting
2. accepts and praises students' responses and expands them
3. refers to the rule chart posted in the room to aid comprehension
4. asks open-ended questions to encourage more language production
5. paraphrases her own sentences to expand vocabulary and assist student comprehension

Advanced Stage

Students at the advanced stage can give you the impression that they are quite fluent in English. They are able to retell events in greater detail, usually use standard grammar, and speak with fewer errors. It is easy to assume that students at this stage are no longer English learners. However, students at this level still need a great deal of support, especially language that is used for content area instruction. These students require substantial practice with academic language in order for them to be successful in school. At this stage, too, we believe it is appropriate to teach grammar as long as it is integrated within meaningful communication activities. Many published programs for teaching English as a second language include instructional activities for teaching grammar. But again, we caution you to avoid drills and exercises that isolate skills from meaningful communication.

Academic language can be described as language that is used for the purpose of acquiring new knowledge and skills (Chamot & O'Mally, 1994). Students need to be able to follow oral explanations of concepts that are taught in the various subject areas. They also need to be able to use appropriate vocabulary to describe how to solve a math problem, explain the conclusions of a science experiment, and describe causes for an historical event. Thus, for students at the advanced stage we recommend that you:

- Continue to focus on vocabulary used in subject-area instruction such as math, science, and history/social science. This involves teaching new words as well as teaching new meanings of words students may already be familiar with (i.e., map terms such as *legend* and *scale*).
- Expose students to more complex language structures. Model the language that is necessary to

understand grade-level concepts and grade-level textbooks.

- Continue to provide comprehensible input by using charts, pictures, providing examples, and/or demonstrating how to convey the meaning of new vocabulary and expressions. Paraphrasing, repeating key vocabulary, and summarizing main points are also effective strategies.
- Constantly monitor students' comprehension by asking questions frequently. Questions can check for understanding of key terms (What is a *carnivore*?) but should also include high level open-ended questions that require students to compare and contrast and analyze and evaluate (How do you think the climate in this region would affect the people's lives?).
- Provide students with many opportunities not only to listen to academic language that is made comprehensible for them but also to *practice using academic language.* For example, through cooperative learning activities in small groups or with partners, students can create a bar graph and explain it to the class or conduct a science experiment and describe their procedure and findings.

Chamot and O'Malley (1994) offer many excellent suggestions for integrating the teaching of content with academic language. More strategies for teaching the content area subjects to English learners will be discussed in Chapter Seven.

Advanced students also need instruction to broaden their vocabulary related to literature. Most language arts programs include many activities to develop vocabulary that are also appropriate for English learners. But keep in mind that English learners will need more explicit instruction and practice with words and expressions with which they have had little experience. Vocabulary development is closely related to reading and academic achievement, so we will discuss this topic further in Chapters Five and Seven.

Important Points to Remember for Advanced Stage

- Focus on vocabulary and academic language used in all subject areas.
- Provide comprehensible input when new terms are introduced.
- Do not assume that students understand; always check for comprehension.
- Provide many opportunities for students to practice using academic language.

Vignette # 3

In the following vignette, can you identify four tools the teacher uses with advanced level fourth grade students to further their language development?

The teacher shows pictures of different birds and explains how different types of beaks are used to gather food.

Teacher: Duck beaks are flat and are used for straining water. *Straining* means to separate small solid pieces of food from the water.

The teacher then demonstrates the process of straining using a kitchen strainer.

Teacher: What am I doing?

Student: Straining.

Teacher: Yes, another word for straining is filtering. How can straining, or filtering, be useful to people as well as birds?

Student: We strain spaghetti.

Student: Yes—and beans.

Later, each student selects a bird to illustrate how it uses its beak to gather food. Then, working in pairs, they explain to a partner how this is done. The teacher walks around the room to listen and provide assistance.

Student: This beak is short and has a hook and it's good for tearing flesh.

Teacher: That's right. It's a short-hooked beak. How does this type of beak help the bird to survive?

Student: They get to eat what they tear up.

Tools

What tools does the teacher use?

1. demonstrates to convey meaning
2. provides for student interaction to practice academic language
3. expands on student responses to model language
4. uses open-ended questions to promote language and thinking

Additional Activities for All Levels

In addition to the tools already mentioned, there are others that can help you support your English learners' oral language development at all levels.

Front-loading vocabulary: Prior to presenting a lesson, preview the vocabulary using pictures and other means to convey meaning. In addition, after the lesson you can review the vocabulary and in this way students will have more opportunities to hear and learn key terms.

Singing: Children enjoy singing and this can be an excellent way to teach vocabulary. Music can reduce anxiety and can be a great motivator for even very shy students to become involved in the activity. The meaning of lyrics can be conveyed through pictures or gestures. Some teachers substitute their own lyrics to popular songs to teach vocabulary and basic concepts. To support word substitution for a familiar song, divide a large piece of butcher paper into sections for nouns, verbs, adverbs, and prepositional phrases. Students can draw from vocabulary they've learned throughout the unit of study. The teacher then guides the students to create sentences that they sing to the familiar tune of *Farmer in the Dell* (Brechtel, 2001). Here's an example to help you. During a unit on the body, your students might create something like: "The dark red blood cells; the dark red blood cells, the dark red blood cells carry oxygen through the body." It will be helpful for you to create a chart with categories of words that are in the same order in which they will be used to make up the chant, in this case, adjective, noun, verb, and prepositional phrase.

Chanting: Chanting involves the repetition of words or phrases using stress and rhythm. You are probably familiar with the traditional jump rope chant *Teddy Bear, Teddy Bear* that can be used for teaching a variety of actions. Books by Carolyn Graham are excellent resources for chants that use jazz rhythms, along with clapping and other actions, to teach the natural stress and intonation patterns of American English. Two of the most popular are *Jazz Chants* and *Jazz Chants for Children* (both are Oxford University Press, 1979). Audio cassettes and CDs are also available to accompany these books.

Rhymes and Poetry: Reciting rhymes and poetry can help students acquire intonation patterns of English. As with songs and chants, accompanying the recitation with actions can help students understand and remember the words.

Dramatization: It is very motivating for children to act out stories or events. Props and pictures can be used to help support meaning and stimulate interest. The use of puppets can also be very helpful in involving children who are hesitant to participate. They may forget to be self-conscious because the puppet is the one speaking! As students become more proficient in the language they can even create their own skits.

Games: This is an excellent vehicle for engaging children in practicing English. There are many different types of games. We will mention only two of them here. You may recognize these since they have become such an integral part of so many classrooms.

> Who Am I Thinking About? A student selects a card with the name of a class member (or a famous person related to a unit of study). The student then describes the person to the class who must guess who it is. This game can be varied to guessing a place or an object.
>
> Draw What I Say: One student has a picture that he or she must describe to a partner. Without looking at the picture, the partner tries to draw it based only on the description.

The range of activities that a teacher can provide to support children's language acquisition is large and limited only by the imagination and resources of the teacher. We encourage you to try out a wide variety of activities that are stimulating, fun, and most importantly, motivate students to communicate in English.

Important Points to Remember

- School-aged English learners need to acquire academic concepts at the same time that they are developing proficiency in English.
- Fluent oral language development for social uses may occur in two to three years while academic language will take longer to develop.
- Comprehensible input as well as many opportunities to use English in social contexts to negotiate for meaning are key components in developing language fluency.

Post-Reading Reflection

1. Identify three strategies you could use to develop English oral language with beginning-level students.

2. Intermediate-level students often make grammatical errors in their speech. How should a teacher respond to such errors?

3. List three things you could do to motivate English learners to communicate in English.

4. Identify three things you could do to help English learners who are at an advanced stage of English language development.

Chapter Five

Teaching Reading

Fluency in spoken English plays a critical role in learning to read in English. Because oral language is such an important factor, it is preferable for children to learn to read first in their native language (Fillmore & Valdez, 1986; Thonis, 1994). We recognize, however, that this may not be possible for many reasons, including lack of reading material, the number of native languages represented, lack of qualified personnel, and/or state and district policies. English learners can be successful in acquiring literacy skills in their second language, and there are many things you can do to promote their achievement. This chapter will compare first and second language reading, describe some of the difficulties that English learners encounter when they are learning to read in English, and provide classroom tools that you can use to facilitate your students' progress in reading.

Differences in the Reading Process Between English Learners and Native Speakers

The teaching of English literacy skills does not need to be postponed until the students are fluent in English or even until they are at an intermediate level. The development of literacy in English can begin quite early in the acquisition process (Crawford, 1994; Fitzgerald & Noblit, 1999). This is particularly true for older students or for those who already have some literacy skills in their native language. The important thing to remember is that *effective reading instruction for English learners involves many of the same strategies as effective reading instruction for native speakers of English.* The process of learning to read in English seems to be very similar for both English learners and native English speakers (Peregoy & Boyle, 2000). All readers learn to use their knowledge of conventions of print, structure of the English language, and background knowledge to make sense of text. However, there are some important differences between native English speakers and English learners. Important factors that contribute to these differences include proficiency in English, background knowledge, and literacy in the first language. A key instructional difference is that *meaning-making and vocabulary must be even more central to the reading process for English learners than for native English speakers.*

Proficiency in English

The first thing to consider is that English language proficiency will influence students' ability to read fluently and, most importantly, to understand what they are reading. This is because, as Goodman (1996) tells us, proficient readers use their knowledge of three linguistic systems or cues to construct meaning when they read. These systems are: (1) the graphophonic system (ability to use letter-sound relationships to decode text); (2) the syntactic system (ability to predict what word is likely to come next in a sentence based on knowledge of word order); and (3) the semantic system (ability to derive meaning of words on the page). Students who are not yet proficient in English are not able to effectively use all of the linguistic systems that good readers use in order to construct meaning from text. Although English learners may learn to decode, they will still struggle with using the other two cueing systems because of their limited English language. Therefore, you will need to emphasize language development in all reading activities. The more familiar English learners become with English syntax and especially vocabulary, the easier reading in English will become for them.

Background Knowledge

Another difference is the prior knowledge that English learners bring to the task of reading in English. English learners come to school with a great deal of knowledge about their culture and communities. However, this may not be relevant to the background knowledge required to understand the stories and textbooks they read in school. If a story centers around an American celebration, such as Thanksgiving with traditional American dishes, a teacher will need to develop some background knowledge with English learners who are recent arrivals from other countries in order to make the story comprehensible to them.

First Language Literacy Level

A third consideration is whether students have some literacy skills in their native language. Knowing how to read in the first language will facilitate learning to read in English (Cummins, 1994). Even if the child is literate in a language that uses a very different writing system than English, such as Hebrew or Arabic, primary language literacy will benefit the development of literacy in English. For one thing, these students already have the crucial understanding that the printed symbols on the page represent speech and carry a message. If the native language writing system is more similar to English, such as Vietnamese or Spanish, then the child already understands directionality concepts about print, such as left to right and top to bottom. In addition, once students have learned comprehension skills, such as identifying a main idea, drawing inferences, or making predictions in their primary language, these skills will transfer to reading in English (Thonis, 1994). Therefore, promoting literacy in the native language, whether at home or at school, is very beneficial to English learners.

Teaching Strategies for Beginning Reading

Since English learners develop literacy skills in similar ways to native English speakers, they will need some of the same instructional strategies. English learners should be immersed in meaningful, purposeful reading and writing activities that teach basic concepts about print and the numerous ways reading and writing are used for communication. At all times, comprehensible input should be provided to convey meaning of the printed word.

It is well-established that the development of phonemic awareness and the teaching of phonics are important in the process of teaching children to read (Adams, 1990). Research also demonstrates that teaching these skills to English learners can benefit their reading development (Chiape, Siegel, & Wade-Woolley, 2002). Phonemic awareness is the understanding that speech is composed of a series of sounds and that those sounds can be isolated and manipulated (Tompkins, 2006). Teachers can develop phonemic awareness in children by providing many activities that involve songs, rhyming, word play books, and word games (Yopp & Yopp, 2000). Phonemic awareness provides a foundation for learning phonics, which is the understanding that speech sounds can be represented by letters and letter patterns (Gunning, 2003). A child will not be able to understand, for example, that the sound of /*t*/ at the beginning of the word *toy* can be represented by the letter *t* unless the child can hear and isolate that sound from the rest of the word.

The ability to recognize high frequency words is also critical to successful reading, so that attention can be focused on comprehending the text and not on decoding each and every word. As with native English speakers, it is essential to emphasize that the main task of reading is to construct meaning. Thus, instruction should always include real reading with text, such as poems, stories, or song lyrics. Prior to focusing on individual sounds, letters, or words, develop children's understanding of the words (Peregoy & Boyle, 2005).

As mentioned earlier, the same strategies that are recommended by reading experts for developing literacy skills with native English speakers are also appropriate for English learners. However, English learners will have some special needs that teachers need to consider:

- *English learners will need to listen to word play books and rhymes many more times than native English speakers.* Some teachers we know have made their own tapes of books and rhymes for English learners to listen to again and again at listening centers as well as at home.
- *When teaching phonemic awareness, make sure students are familiar with what the words mean before they are asked to identify and manipulate sounds in the words.* Use pictures and concrete objects to convey meaning and make activities as interesting and engaging as possible.

- *English learners, especially at the beginning level, may not be able to perceive or reproduce accurately some English speech sounds.* Do not stress correct pronunciation—this will only create anxiety. (Imagine yourself quickly learning to correctly pronounce words in a foreign language.) Given continued exposure and practice with English, perception of sounds and pronunciation will improve. The important thing is that the student understand the meanings of the words.
- *Always combine the teaching of skills with exposure to meaningful text.* Instead of drills and activity sheets, use children's literature or create sentences and stories that incorporate the children's names and interests to provide practice with identifying letters and sounds.
- *Teach high frequency and sight words in context.* Some of these words such as *an* or *of* can be very difficult for English learners to understand unless they are placed in meaningful sentences. First, teach the words within sentences; then, once the children become familiar with what they mean, the words can be isolated for reading practice. Students can create their own flash cards and illustrate the meanings on the back of the cards.
- *Reading aloud to students benefits English learners in particular because the teacher may be the only model of fluent reading and intonation patterns in English.* Select a wide variety of texts that are appropriate for the age and interests of your students and point to the words as you read them. Use pictures in the books, gestures, or pantomime to convey meaning. Repeated reading aloud of familiar books, accompanied by comprehensible input, can be very enjoyable for children, while also contributing to their acquisition of English.

- *Engage your English learners in many shared reading activities.* Shared reading involves students more directly in reading while providing important support to ensure their success. Using books that the students cannot read independently, the teacher or a more skilled reading buddy does most of the reading but invites English learners to join in with reading repeated words and refrains. English learners should have many opportunities to see and hear the text so they can pick up repeated words and phrases and read along successfully. This works well for all grade levels and is an excellent way for students to gain confidence with reading.
- *Use Reader's Theater as a way for students to practice reading.* In this approach, students assume the roles of characters in a story and practice reading their parts using gestures and appropriate expression to interpret the character until they are ready to perform in front of the class. Even beginning-level students can often participate in a Reader's Theater presentation if their lines are limited and they have ample time to practice. Adding a few simple props helps to support meaning and can add interest and enjoyment. You could also use the buddy system—pair an English learner with a skilled reader to practice and perform the lines together.
- *Use the Language Experience Approach.* This approach has long been recommended for learners of all ages to teach both reading and writing skills and to initiate English learners into English print (Crawford, 1994; Law & Eckes, 2000). The students first participate in a shared experience to generate language. This can be listening to a story, going on a field trip, attending a school assembly, or participating in a class event. The students then

engage in discussing the event and dictating phrases or sentences to develop a story about it, which the teacher records on chart paper in front of the students. The students read their own words back to the teacher, and the teacher tracks the print as the students read. This approach is very effective for modeling the reading and writing process, teaching relationships of print to speech, teaching sight words, and giving instruction on many other concepts about print. The language comes directly from the students' shared experience and background so it is of high interest and they are more likely to experience success reading and working with the text. English learners at the beginning levels of oral language development may only be able to dictate single words and phrases. They can copy and illustrate these to make individual books. As students acquire more English, they will be able to dictate more sentences, retell familiar stories, and create their own stories.

Developing Comprehension

Although English learners can learn letter-sound relationships and can become good decoders of text, they often tend to struggle with reading comprehension because they lack the vocabulary and background knowledge needed to make sense of the text. It is important to keep in mind that in most cases when English learners struggle with the task of reading in English, it is because they have not had the same years of experience with English as native English speakers have. The native speaker of English has five years of English language development prior to starting school, which is a considerable head start on English learners. Moreover, native English speakers continue to expand their vocabulary and language skills every year. Thus, as Cummins points out, English learners "must catch up with a moving target" (1994, p.14). The good news is that there are many proven instructional strategies teachers can use to effectively teach vocabulary to English learners. Furthermore, extensive reading has the largest impact on children's vocabulary growth (Tompkins, 2006). For Evelyn, who grew up in a Spanish-speaking home, becoming an avid reader in elementary school provided her with knowledge of new English words on a regular basis. The critical task for teachers, then, is not only to provide English learners with direct instruction in vocabulary but also to motivate them to read often and independently. Instilling a love for reading is possibly the single greatest contribution that you can make to your English learners' success in school.

Here are some suggested strategies for supporting English learners' comprehension skills:

- *Before reading a story, select words and concepts that are key to understanding it.* Aim for no more than eight or ten. Use pictures, gestures, concrete objects, and actions to convey the meaning of this vocabulary to prepare students to understand these in the story. Dramatizing the meaning of words can be particularly effective with English Learners because the physical action can help them to remember the words.
- *As students advance in their English, more difficult words can be explained and discussed.* However, it is still important to provide multiple opportunities for vocabulary development. Gunning (2003) suggests that teachers:

    1. present words within the context of the story
    2. provide an understandable definition

3. have students compose sentences to relate the new words to their lives
4. review words regularly and encourage students to use them

Repetition of vocabulary words in a variety of interesting activities is important. Students need to hear and practice saying new words many times before they are internalized.

- *Create Word Walls.* These are recommended by many reading experts as a means to support students' vocabulary development as well as sight vocabulary (Cunningham, 2000; Tompkins, 2006). For English Learners, word walls can be enhanced with pictures or children's drawings to help illustrate meanings. To create a word wall, use large pieces of chart paper that are divided into sections, one for each letter of the alphabet. Place words on the charts as they come up in books, songs, poems, or other activities. It is recommended that two different types of word walls be used—one for high frequency words and another for vocabulary words that arise in stories or content instruction (Tompkins, 2006). Word walls can serve as a useful tool for students' writing, which we will discuss in Chapter Six.
- *Have students make their own individual dictionaries in which they record words, illustrate them, and, depending on their abilities, write sentences using the words.* Student-made dictionaries can be helpful resources to which students can refer when they are writing.
- *Use visual displays, such as charts, diagrams, and webs to help students see relationships between words and ideas from literature.* For example, charts can be divided into sections for listing nouns, verbs, and adjectives or to compare

synonyms and antonyms. A common type of web has a main idea or word in a center circle with lines that radiate out from the center. Related words or ideas are then written at the end of each of the extended lines. Another useful visual is the Venn Diagram, which consists of two overlapping circles that can be used to list differences and similarities between two objects, living things, stories, or something else the children may want to compare. (See Appendix)

- *Use objects that students can touch and handle.* For example, you can present students with a box containing various objects related to a story, such as toy animals representing some of the characters in *Charlotte's Web* by E.B. White (Harper Trophy, 1974). This "Book Box" activity can be used to teach vocabulary as well as to stimulate interest in the story and engage students in making predictions about it (Yopp & Yopp, 2001).
- *Provide concrete experiences that can help students acquire needed background knowledge to understand a story that is read.* Although field trips and videos are helpful, sometimes these are not possible or necessary. Something as simple as bringing in pictures or objects or having a short discussion about an unfamiliar topic can be useful. For example, before reading *Corduroy* by Don Freeman (Puffin, 1976), bring in a piece of corduroy fabric for the children to see and touch. Or before reading *Ira Sleeps Over* by Bernard Waber (Houghton Mifflin, 1975), discuss the custom of children staying overnight at a friend's house, which may not be familiar to many immigrant children. In addition, whenever possible, include reading material that reflects students' cultural background. Many are available on the market today.

- *Group your English learners with native English speakers whenever possible, because native speakers can provide good language models for those still developing English.* Social interaction fosters language acquisition; therefore, provide many opportunities for students to work in pairs or small groups to discuss elements of stories or work on literacy tasks (making books, creating charts, etc.).
- *Teach students to draw on cognates that they already know in their native language to help them understand English words. Cognates* are words that are similar in two languages and have the same meaning but may be pronounced differently. Spanish and English have many cognates (*artista/artist, círculo/circle*) that Spanish speakers can use to help them unlock meanings of unknown words they encounter.

Vignette # 4

Read the following classroom vignette. Can you identify four tools that the teacher uses in the following vignette?

> Christina wants to teach her first grade English learners the long /e/ sound of the letter patterns *ee* and *ea.* One of the books that Christina has read aloud to her students many times is *Sheep in a Jeep* by Nancy Shaw (Houghton Mifflin, 1986). Before reading the book, Christina showed the children a toy jeep and pictures of sheep and hills. Christina talked about and demonstrated the meaning of *steep* using gestures and the picture of the hills. While reading the book aloud, Christina used the book's illustrations and gestures to demonstrate the meaning of words such as *shove, tug,* and *weep.* Christina presents this by now very familiar book to the children and asks them to listen for the long /e/ sound. The children are asked to place their hands on their heads when they hear the sound in a word that is read. As she reads the book, Christina emphasizes words like *sheep, jeep, beep, cheap, leap,* and *weep* and checks to see that all the children are responding appropriately. After reading the book, Christina shows prepared sentence strips from the story that contain these words. Together they read the sentences, and Christina highlights the letter patterns of *ee* and *ea* that represent the long /e/ sound. Christina then guides the children to generate other words they may know that have this sound, and she records them on a chart. When some of the Spanish-speakers confuse the *ch* and *sh* sounds, saying *cheep* when they mean *sheep,* Christina models the correct pronunciation and checks for comprehension (Do you mean *sh*eep, the animal?) Later the students read another story and, working in pairs, they look for words that contain the *ee* or *ea* pattern.

Tools

What tools does the teacher use?

1. uses objects, gestures, and pictures to convey the meaning of vocabulary
2. combines the teaching of skills with meaningful text
3. focuses on student understanding, not on correct pronunciation of sounds
4. provides an opportunity for students to work together

Vignette # 5

Can you identify five tools that the teacher uses in the following vignette?

> Diana has many English learners in her fifth-grade classroom. The students had completed reading an article about stingrays. She is reviewing portions of the text and questioning the students on their understanding.
>
> "If a stingray sees you coming, is it going to come at you and attack you?" asks Diana.
>
> "Not if you don't bother him," responds Gabe.
>
> "Right, the fourth line says, 'They will not go out of their way to attack you.' That means they are not going to think, 'I'm going to get him.' (She uses a pinching hand movement.)
>
> The stingray article points out that when wading, it is wise to be alert and to prod a stingray with a stick to make it move away. To reinforce this point, Diana engages the students in a very brief role play of two "waders" and a "stingray." She asks the "stingray" to place himself on the floor to "hide" under the sand.
>
> "O.K. Gabriel. You're our wader. Now you're wading in the water. Do one of two things the text tells us. What are you going to do that the passage says? What's the word the text uses?"
>
> Several students in the class respond, "It says *prod*."
>
> Diana nods and asks, "And what's another word for *prod*?"
>
> Several students respond, "*Poke*."
>
> Gabriel "wades" a few steps then he pretends to prod the "stingray" with a stick. The "stingray" scuttles away.
>
> "O.K. Susana. You're our next "wader." What are you going to do?"
>
> Susana responds, "I'm going to step it."
>
> "Yes, you're going to step on it," says Diana.
>
> Susana "steps" on the "stingray," and it lashes out and stings her.
>
> Diana then asks the students to add two new vocabulary words and their definitions to their personal dictionaries. (Zuñiga & Yopp, 1996)

Tools

What tools does the teacher use?

1. provides many opportunities for vocabulary development
2. uses real-world connections
3. creates a dramatization to make text comprehensible
4. asks students to add new vocabulary words to their own dictionaries
5. groups English learners and native English speakers together

Important Points to Remember

- English Learners at the beginning stage can be successful in learning English reading skills, but this must be accompanied by comprehensible input, and skills must always be taught within meaningful context. At all times, focus on reading for meaning.
- Reading experts stress the importance of providing daily opportunities for reading to students, having students read with others, and having students read independently (Gunning, 2003; Tompkins, 2006). This is particularly essential for English learners who require lots of repetitions. So, read and reread books many times. Children never tire of familiar engaging stories.
- Knowing how to read in the native language can help support reading in the second language, so encourage parents of English learners to read to their children in their native language. Often, teachers believe that parents who do not speak English cannot do very much to help their children be successful in school. Not true! Not only can they help by promoting literacy in the native language, but they can also convey the importance of reading, provide a quiet place for study, and communicate support and love for their child.

Post-Reading Reflection

1. List three things you can do to help English learners develop their reading skills in English.

    ____

    ____

    ____

2. Why is it that English learners may quickly learn to decode but often struggle with comprehension?

    ____

    ____

    ____

    ____

3. List three ways you could help English learners improve in reading comprehension.

    ____

    ____

    ____

4. How can the ability to read in their native language help English learners develop reading skills in English?

    ____

    ____

    ____

    ____

Chapter Six

Writing

Let's begin with what may be a review for you on some important facts that we know about writing.

Ten Truths About Writing

1. Just as with oral language, writing develops in stages. The more writers practice and the more support they have, the better they get.
2. All writers must consider speaker, audience, and purpose. Who are you? To whom are you writing? What's the message that you are conveying? A writer needs to be clear about these elements in shaping a piece of writing.
3. Writers select interesting topics to write about, authentic reasons to write, and have regular opportunities to write (Temple, Ogle, Crawford & Freppon, 2005). Writers select topics based on their interests. They want to convey something about a topic of interest and are probably not so eager to write about an assigned topic. They have a reason for writing about a specific topic—it

could be to share a passion, to find out more through writing about it, or a number of other reasons. Internal reasons for writing are more compelling than external ones. Writers need to have opportunities to write regularly. Writing is creative but it is also disciplined. The discipline of sitting down to write at a regular time helps develop the creativity and the expectation that one will in fact write.

4. Writers go through the process of brainstorming or prewriting, drafting, revising, editing, and publishing. While publishing is the final stage, the other stages do not necessarily occur in linear order. For example, a writer can brainstorm, then write a draft, then brainstorm some more, then revise a portion, then revise the draft, etc. However, for young writers, the writing process is a helpful one to stay close to. It provides a good mental model to follow.

5. The four language domains of listening, speaking, reading, and writing are integrated. Development in one supports development in another.

6. Writing can be the most challenging of the four domains of language. Why? Listening and reading are receptive language skills. Speaking and writing are productive language skills. First, it takes more effort to produce than to receive. Second, putting something in writing can be intimidating. It's there for everyone to see—mistakes and all. Collins (1998) says that writing is a secondary form of discourse. Speech, he suggests, is the primary source of communication and something we do naturally. Writing, then, is one step removed from what comes naturally to us. Humans do not need instruction on how to learn to speak. We do, however, need instruction on how to learn to write.

7. Students must develop the ability to become as fluent as possible in four domains of writing—poetry, narrative, expository, and persuasive writing. School curriculum typically requires direct instruction on helping students develop some level of competence in each of these domains.
8. Every writer benefits from assistance. Writers' workshops tend to be more open-ended and collaborative. Mini-lessons that teachers do in classrooms provide explicit direction or instruction. Mini-lessons model, instruct, or demonstrate specific points.
9. Writers actively pay attention to other people's writing. We become more aware of good writing and what makes good writing. We look for strategies that we're familiar with and begin to notice strategies that we're not familiar with. It is very beneficial for writers to read and hear other authors' writings.
10. All writers want at least some of their work published. For the classroom, this could mean a clean, final edit that you bind in some way, to give students the feel of a published book. As a classroom teacher, be certain to give this opportunity to your students. It is deeply satisfying for writers to see their work in print. It is a source of satisfaction and pride. It also serves as a stimulus for further writing.

How is Writing Different for English Learners?

1. *You'll need to consider the student's level of oral English proficiency as a guide for what level of writing to expect.* You can use oral language

development (e.g., the SOLOM discussed in Chapter Three) to help guide your expectations.

2. You will need to consider the student's native language writing proficiency and level of formal schooling. *The more advanced the student's writing in the native language, the more knowledge about writing he/she will be able to transfer to writing in English.* For example, look at this sentence written by a fourth grade student:

   –"What is the matter?–" she said.

   You may notice that he included both the English (quotation marks) and Spanish (dashes) forms of indicating speech. This student already understands that writers indicate direct speech. He just requires a bit of fine-tuning on the use of quotation marks in English rather than direct instruction and practice on the use of quotation marks.

3. Support and instruction are important for all writers; however, *models, patterns, structure, organization, scaffolds, and direct guidance become even more important for helping English learners develop good writing skills.*

4. *English learners cannot rely on native speaker intuition or sense of language* (De Jong & Harper, 2005). Native English writers have an innate sense that tells them when something is not correct or does not flow well. More specific types of feedback are necessary for English learners to help them discern when a piece of writing is not quite right.

5. *Teachers who do writing conferences with students will need a tighter focus.* For example, it may be an appropriate moment to do an ESL lesson on a specific point of grammar or direct instruction on semantic nuances of a word a student has written. (Example: "I like this lovely

story. It let me feel so sweet when I read it.") (Law & Eckes, 1995). It is very important to limit the purpose of a writing conference to one or two points. An English learner can quickly become overwhelmed and discouraged with rewriting, editing, or grammar instruction overload.

6. As with reading, *English learners will need a great deal of assistance with vocabulary.* A visually print-rich environment—word walls, charts, unit vocabulary, high frequency words—can help. Prewriting brainstorming is another very good tool for generating lots of vocabulary.
7. Mastering writing in a second language takes effort and time. You may recall the discussion from Chapter Two concerning the length of time it takes to become literate—from five to eight years. *For most English learners, writing is the language domain that is the last one to be perfected.* Recall from Chapter One the idea that language learning takes time. Your guidance and encouragement are very important to your student's success.

Writing Strategies Especially Useful for English Learners

We've developed what we think is a useful chart on page 111 (Table 6.1) that will help you guide your English learners in developing their writing skills. Down the left column of the chart, you see the domains of writing that students should develop and across the top row of the chart, you see the three stages of English language development. In each coordinate of the chart, there are very brief descriptors of the types of writing activities or instructional tools that can be useful in that particular domain. In the pages following Table 6.1, you will find explanations of each descriptor—what it is, how to do it, and why it's effective.

We want to be clear that the explanations of these instructional tools are brief. A good reading/language arts methods text will expand on our explanations. You may find those that we've referenced in this book especially helpful. Also, be mindful that some of these writing activities take several days to accomplish with your students. Others require you to repeatedly model with your students. Our primary purpose here is to provide helpful teaching tools for you to use with your students at the three stages of writing development.

Table 6.1: Writing Activities for English Learners in Four Domains of Writing

	Beginning	Intermediate	Advanced
Narrative	1. Wordless picture books 2. Story captions 3. Pattern books 4. Lists 5. Greeting cards & postcards	1. Show not tell 2. Sentence combining 3. Story maps 4. Retelling familiar stories 5. Buddy journals	1. Literature response journals 2. Hotseating/reader response
Expository	6. Life murals 7. Lists and maps 8. Time lines and cycles 9. Concept books 10. Language experience approach 11. Found poetry	6. Sentence Patterns 7. Writing structured paragraphs 8. Graphic organizers 9. Content area big books 10. Retellings	3. Sentence patterns 4. Double-entry journals
Poetry	12. Pattern poetry		
Persuasive		11. What-so what-now what?	

Beginning Writers

Basic language patterns and a great deal of vocabulary assistance are especially important at this level. Remember to make use of your older English learners who are at the beginning stage of writing as cross-age helpers. They can read their own material to students in the early grades. Pattern books and concept books are especially good for cross-age helpers to share.

1. Wordless picture books

Wordless picture books are excellent for providing writing scaffolds. First, much of the vocabulary is evident in the pictures. Second, the structure or story line is also provided. Students can first orally discuss the story, generating language and ideas. This oral language then leads to writing, such as creating simple sentences that describe the existing story.

2. Story captions

This involves writing story captions for a familiar story. After multiple readings, you can photocopy pictures from the book you've been reading with your students. They can then sequence them and write simple sentences that describe the story. Let's take *The Three Little Pigs* as an example. After hearing the story several times and discussing it, the students will be ready to create captions, such as "This pig used straw," "This pig used sticks," and "The wolf is mean and hungry."

3. Pattern books

Books like *Brown Bear, Brown Bear, What Do You See?* by Bill Martin, Jr. (Harcourt Brace & Co., 1967) provide patterns that, once learned, students can use to create their own oral and written language. Students can use the pattern *I see a* _________ *looking at me*, or they can even add a different verb to create their own stories, for example, *I see a* _________ *smiling at me* or *I see a* ________ *waving at me*, etc.

4. Lists

Students can create lists with categories of words like verbs, adjectives, or prepositional phrases and use them in patterned sentences. For example, you can set them up to use patterns like:

- *You're my friend because you're (adjective).*
  Example: *You're my friend because you're fun.*
- *I love you because you're (adjective).*
  Example: *I love you because you're caring.*
- *I'm happy when I'm (verb+ing).*
  Example: *I'm happy when I'm playing.*
- *I'm happy when I'm (prepositional phrase).*
  Example: *I'm happy when I'm with my friend.*

Students can use this language to create their own simple stories or books.

5. Greeting cards and postcards

Students can write brief messages using familiar words and phrases to create greeting cards that they can send to friends and family members.

6. Life murals

Students can write captions that accompany their illustrations of a single important event that occurred, a day in their life, or their autobiography. The scaffolds provided here are their own memories. They will seek phrases that describe their memories and knowledge of these events.

7. Lists and maps

It is common for beginning English learners to learn new words and phrases based around themes, such as fruits and vegetables, animals, and clothing. Students can create lists of these and make maps to use these words. For example, how would they organize a grocery store, arrange animals in a zoo, or set up a clothing store? Some content areas lend themselves to this type of activity as well. An example is the study of an ancient civilization. Using the words they learn, students can make maps of the locations and descriptions of river systems and physical settings that supported early settlements and civilizations. They can draw maps of principal rivers, showing where products came from and where they were transported to in support of trade.

8. Time lines and cycles

Students can use words and phrases to describe a series of events, such as historical events that lend themselves well to sequencing. Also, as with the story captions activity above, pictures can be photocopied from textbooks for students to use to create time lines with captions. Cycles also lend themselves to labeling and describing with brief captions. Some events in science can be diagrammed in cycles. Examples are the life cycle, the circulatory system, food chains, and the rock cycle.

9. Concept books

These are useful in a variety of ways. (Peregoy & Boyle, 2005). Students can make parts of speech books using verbs (*The book of -ing)* or prepositional phrases ("in my desk," for example). Peek-a-boo books can be made by pasting in flaps of paper that conceal a picture showing "in the drawer" or "on the desk," etc. Other concept books can be used to demonstrate opposites or comparisons (*-er*, *-est*). ABC books can be used in connection with a unit of study, for example, the ABCs of the Westward Movement.

10. Language experience approach

Described fully in Chapter Five, this involves engaging in a shared experience, discussing it, writing about it, and reading it back.

11. Found poetry

This writing activity should take place at the completion of a unit of study. This activity may take several days to complete. It may be more productive to include intermediate-level writers in this activity. Provide your students with a piece of text. This might be taken from their textbooks or another source. The text should include familiar vocabulary. Copy the text onto an overhead and read the text with the students. Ask students to read words or phrases that draw their attention. These could be words they know or that they find unusual or interesting in some way. Highlight these on the overhead. Provide students with strips of paper onto which they can copy their words and phrases. Have students place them in a pocket chart. Read through all the words and phrases. Have the students arrange these words and phrases in some way that sounds pleasing to them. This will occur through group negotiation. The group must agree on the final arrangement of the words and phrases that make up the poem. You and your students will be delighted with the beauty of the found poetry they create. It is also an excellent way to review familiar vocabulary and concepts.

12. Pattern poetry

Numerous poetry patterns are available and appropriate for beginning writers (Peregoy and Boyle, 2005). Peregoy and Boyle point out that *Wishes, Lies, and Dreams:*

Teaching Children to Write Poetry (Koch, 1970) is an excellent source. One of our favorite teachers, Karyn Mazo-Calf, provided us with guidelines for writing poetry with students. Start slowly and build up, providing lots of examples for students. Take the time to help students edit their own and each other's work. Help creative and imaginative thinking grow by spending ample time in the prewriting phase.

Here are a few ideas of poem patterns from Karyn.

Five line sensory poems:

Love looks like _________.
Love sounds like _________.
Love feels like _________.
Love smells like _________.
Love tastes like _________.

Five line metaphor poem based on a single idea:

Happiness is (a sunset).
Happiness is (my mom's perfume).
Happiness is (the beach).
Happiness is (cookies after school).
Happiness is (slides and swings).

You can also make up your own poem patterns. Here are just a few:

I used to be a _____ but now I'm _________.
I like to _________.
I like to _________.
I like to _________.
But I really love to _________.

Cinquains are also appropriate and fun. They are written in the form of a tree. Here's the pattern for a cinquain:

One-word topic
Two adjectives
Three verbs ending with *-ing*
Four-word phrase
One word that is a synonym for the first word.

A diamante is another poetry pattern. It is written in the shape of a diamond.

Noun
Two adjectives
Three verbs ending with *-ing*
Four adjectives
Three verbs ending with *-ing*
Two adjectives
Noun

Intermediate Writers

Intermediate writers will benefit from scaffolds, models, and writing support that help them expand their vocabulary and language development.

1. Show not tell

This instructional strategy will help your English learners develop vocabulary and descriptive writing skills (Peregoy & Boyle, 2005; California Department of Education, 1996). First, share with your students a paragraph that is rich in description and paints a picture in their minds. Children's literature is full of such language. Here is an example we created:

> *Whispers of creatures are all around blowing their foul breaths on me. Groans and the distant howls of animals fill the air. What's crawling around my feet and legs? Hanging cobwebs*

brush my face. A deep coldness soaks through my coat. Every path ends where it began. I am scared and lost in the witch's enchanted forest.

Compare this paragraph to: "I was scared and lost in the witch's forest."

Showing and not telling helps your students learn to paint a picture with words. Repeated lessons with your students that help them begin to show and not tell will strengthen their vocabulary and narrative writing. Starting with a brainstorm on relevant vocabulary is helpful. Here are a few "tell" sentences you can start with:

She didn't have any friends.

I love my mom.

Family picnics are so much fun.

My dog is old.

This was the scariest moment of my life!

2. Sentence combining

Many intermediate-level students use the basic sentence pattern of subject-verb-object. They need help becoming familiar with and practicing other patterns. You can create several choppy subject-verb sentences around a topic and provide examples of how the ideas can be combined into more complex and interesting sentences. Then provide the students with another set of short sentences. Working in small groups, students combine the ideas and then each small group shares their work with the whole group. You can also give limits, such as writing a maximum of two or three sentences that include all the information from the sentences you gave them. Here is an example of a group of choppy subject-verb sentences you could ask your students to improve upon by combining.

I live on an island. The island is big. The island has lots of plants and animals. Plants are everywhere. Animals walk around all over. There are not very many people. I like this island.

3. Story maps

These are graphic representations of the organization of a story. They give a verbal and mental model to the structure of the story. You can help your students become comfortable with using story maps by first creating some with your students that are based on familiar stories. Providing models is an important part of helping your students to become accustomed to using story maps. These are very effective with English learners because, along with creating the structure of a story, you can help your students add the vocabulary to be included in advance of actually writing the story. It is of great help to have the words and ideas handy in written form, rather than having to search for them at a mid-point in the writing process.

4. Retelling familiar stories

Retelling known stories such as family stories or events provides the scaffold of familiarity. Students know the story well, and so the learning for them is the vocabulary that they need in order to write the story in English.

5. Buddy journals

This is a type of journal maintained between two students who write back and forth to each other. They are fun for paired English learners or paired English learners and fluent English writers who can model standard written English and expand vocabulary. Buddy journals provide a peer audience and an authentic reason to write and read. They give immediate feedback. Students generate their own topics of interest, describe events, share opinions, ask questions, and get to know each other (Bromley, 1995b).

6. Sentence patterns

Expository writing often contains patterns such as "because (of a fact or event), then (another fact)," "if . . . then . . ." and "When . . . then . . ." First, pointing out these types of construction in students' textbooks and discussing how information is organized helps their comprehension. Second, you can then provide instruction and practice on how to use these in their own writing—both in narrative and expository writing (Peregoy & Boyle, 2005).

7. Writing structured paragraphs

Expository writing is structured in a way that narrative writing is not. Teach your students to use a "hamburger" approach to writing informative paragraphs. Imagine a hamburger with several layers—lettuce, tomatoes, onions, etc.—sandwiched between the top and bottom of the bun. Here are the "layers":

Topic sentence
Supporting sentence
Supporting sentence
Supporting sentence
Concluding sentence

Writing sentences on slips of paper and letting students organize them, then pasting them on a sheet of paper, is one approach that works well. Structured paragraph writing is a helpful scaffold to help your English learners write expository text.

8. Graphic organizers

Helping your students "see" and then use various types of text organization is another important scaffold for your students. As we've said earlier, graphic organizers are visual illustrations of pieces of text. They provide useful "maps" for writing. There are several models of graphic organizers. We've included some in the appendix. Remember that you must coach students in their use. You must provide repeated opportunities for your students to both identify them in their textbooks and then to use them in their own writing.

9. Content area big books

A post-unit activity is to create a group big book. Structure is necessary and can be provided in a variety of ways. You can provide scaffolds in a number of ways, too. One way is to answer "want to know" questions from a Know-Want-Learn chart that the class compiles in advance of the unit. Another way is to have each student contribute the most interesting or important fact he/she learned. You can also take "The Important Thing" pattern (see Chapter Three) to provide a writing scaffold.

10. Retellings

Having your English learners first read expository text and then retell it in written form provides a scaffold for their writing. Additionally, with repeated practice, writing a retelling improves text comprehension and helps them improve their expository writing skills (see Chapter Three).

11. What-so what-now what?

Donna Ogle's well-known and widely-used Know-Want-Learn chart serves as the basis for this modification (Temple et al., 2005). You'll construct for your students a chart divided into three columns and label the columns with *What?*, *So What?*, and *Now What?* Ask your students to identify a problem in the "What?" column ("The playground is littered.") Then brainstorm responses with them to "So What?" ("It has germs." "We get sick." "It's embarrassing.") Then ask them to move to possible solutions in the "Now What?" column. ("We should clean it up." "We need a monitor.") This information will provide a structure and vocabulary for persuasive writing.

Advanced Writers

You will find that there is much overlap between writing activities and support that benefit both your advanced writers and fluent English students. As well, you can draw on some of the suggestions that we've given you to support your intermediate-level students. Below we offer you a few more suggestions that provide writing support for your advanced students as they continue to expand their vocabulary and refine their written language.

1. Literature response journals

Even students at the advanced level may find it difficult to respond to literature in completely unstructured ways. A modification is to organize students into groups whereby each one has a specific responsibility. For example, these roles can be *word hunter, event analyzer, character analyzer, connector*, and *questioner*. For each chapter in the book, the students maintain their responsibilities. The word hunter brings definitions of *x* number of words the group might have found difficult—no more than ten. The event analyzer searchers for the key event(s) that happened. The character analyzer focuses on the important things about the character(s). The connector works on making a connection in some way to the students' lives. The questioner brings up important questions about this chapter. The students keep literature logs as they read and share with each other. This organization divides focus and responsibility, and it provides opportunities for student discussion. One of our graduate students did a study using this instructional tool with her sixth grade English learners and—to her amazement—found that it is a very effective instructional tool. She also found that her students held each other accountable for their responsibilities!

2. Hotseating/reader response

This instructional tool works especially well with older students. The whole class or group reads the same piece of literature. They then write questions they would like to ask one of the characters. Then, you form small groups, and each group studies one character and prepares one individual to sit in the "hot seat" to answer questions from the whole group. The students then write a reflection about a particular character or the story line and make connections to their own lives and circumstances. The depth of analysis and reflection that students write about after hotseating will surprise you (Ogulnick, Shelton-Colangelo, & Williams, 1998).

3. Sentence patterns

As with intermediate writers, you should continue to point out sentence construction and words that will help your students' comprehension and that they can use in their writing. Words like "moreover," "nevertheless," "however," " notwithstanding," " additionally," and "although" are examples.

4. Double-entry journals

Both the left and right sides of the brain are involved in processing information with double-entry journals. First, students brainstorm on the left side of the page, asking questions and writing what they know or drawing pictures about the topic. The idea is to generate interest and activate prior knowledge the students may have. Immediately after instruction and reading, students write down what they remember from the lesson.

Students will combine new knowledge with what they previously knew or answer questions they posed prior to instruction (Ruddell, 2006).

Suggestions For Assessing Writing

Writing Portfolio

We urge you and your English learners to develop and maintain a writing portfolio. This will allow you and your students to track development over time. Keep writing samples in each of the writing domains and in each content area. We've developed a simple form you may want to use (Figure 5.2).

Figure 5.2: Assessment of Writing

Date	Writing sample being assessed	Observed strengths	Areas to work on	Instructional strategies	Comments

Finally, we've developed a Holistic Scoring Rubric for English Learners for you to use in assessing your English learners' writing. This may be useful for you if you have access only to a rubric developed for fluent English writers. In each descriptor, we've considered fluency, form, and correctness—the three elements that good writing encompasses.

Holistic Scoring Rubric for English Learners

1	Great difficulty in expressing self One or two ideas only Rudimentary vocabulary Difficult to understand Sentence fragments or phrases Mechanics are rudimentary or lacking
2	Idea or intent is conveyed, although incomplete or in very brief sentences Discernable sentence or two Little or no support for statements or ideas Vocabulary is weak or limited Poor grammar and/or mechanics
3	Stays on topic but content is meager May have choppy or run-on sentences and/or no sentence variety Vocabulary is not extensive May lack paragraphs Moderate control of grammar and spelling
4	Topic development evident Exhibits some support/details May use some sentence variety Good vocabulary Adequate mechanics, although not perfect Has closure
5	Substantial content and support for ideas Good variety in vocabulary Uses some variety of sentence structure Very good control of grammar and mechanics Some transitions evident
6	Conveys ideas very effectively Content is very well developed Writing flows well Ideas are clear and well supported Nearly standard grammar, spelling, and usage Descriptive language Narrative writing demonstrates originality; creative flair Uses a variety of sentence structures effectively Very good sense of mechanics and paragraph development Closely approximates native English writing

Vignette # 6

Read the following vignette. Can you identify six tools the teacher uses?

> Karyn's third graders were doing a literature unit on *Tom Thumb*. The students were discussing the giant's bad mood. Karyn expanded the discussion by asking the students to think and then talk about their own reasons for good moods and bad moods. As the students provided reasons, she began a T-chart on the board of phrases the students generated. She then asked the students to write poems about their good and bad moods. She wrote this formula on the board for them to use:
>
> I get in a bad mood when ________.
>
> But ________ makes me smile.
>
> Each of the students wrote several stanzas, selected their favorites, and then Karyn created a class book titled *Bad Moods, Good Moods* (Zuñiga & Yopp, 1996).

Tools

What tools does the teacher use?

1. uses literature to springboard into a writing activity
2. asks students to draw from their own experiences for writing material
3. helps students produce a great deal of vocabulary before she asks them to write
4. writes the words for students to use in their own writing
5. provides a pattern to scaffold their own writing
6. publishes her students' self-selected best work

Important Points to Remember

- Models, patterns, and scaffolds are especially important to support English learners.
- You will need to give special attention to helping your students' vocabulary development.
- You will need to be specific about helping your students develop knowledge about the structure of written English.
- For almost everyone, writing is the last language domain to be fully developed.
- Providing your students with consistent opportunities to write is the best way to give them opportunities to improve.

Post-Reading Reflection

1. How is learning to write in English different for native English speakers and English learners?

2. Can you describe ways to engage your English learners in expository and narrative writing at all stages of language development?

Chapter Seven

Content Instruction

In order for English learners to be successful in school, they need to understand concepts and be able to use the language of academic subject areas. High quality content instruction that students can understand is critical. How well students understand and learn subject matter is a much better indicator of academic success, rather than how proficient they are in English (Callahan, 2005). Although understanding new concepts in content areas is cognitively demanding, you can use a variety of strategies to help English learners master complex material. In this chapter, we will discuss the various ways in which you can modify content instruction so that it is more comprehensible for English learners.

Sheltered instruction refers to instruction that uses techniques to make content accessible or that makes language and concepts *visible* to English learners (De Jong & Harper, 2005). This approach incorporates the use of visuals, modifications in teacher talk, many opportunities for interaction, and many principles of good teaching. It is often said that *sheltered instruction* is simply "good teaching." We would agree that it basically *is* good teaching, and native English speakers can also benefit from this type of instruction. However, English learners will require substantially more comprehensible input than will native English speakers. English learners will also need more support to understand subject matter textbooks. Moreover, the purpose of sheltered instruction is twofold: (1) to teach content and (2) to teach academic language. These are important because English learners may have limited exposure to academic language outside of school.

English learners at the advanced level will benefit the most from sheltered instruction. Beginning and intermediate-level students can and should participate in sheltered instruction; however, they will not be able to fully understand complex and abstract concepts until their language skills become more developed. This is especially true for students in the upper grades, when learning becomes more cognitively demanding and complex. For this reason, we recommend that beginning- and intermediate-level students be provided with support in their native language if at all possible. Note that including them in sheltered instruction is not detrimental, however, they may not benefit as much as the more advanced students. Less linguistically advanced students can acquire a great deal of language through content area instruction. Additionally, they will grow in self-esteem as they become full participants in your classroom.

If your English learners are all at the same level of English proficiency, you will spend less time planning because the language that you use, the questions you ask, and the tasks that you assign will all be designed for just one level. It is more likely, however, that your students will be at varying levels of language proficiency. Consequently, you will need to think about different objectives, assignments, and student responses based on these different levels. This does not mean that you need to plan separate lessons for each student level. Rather, include multilevel objectives, assignments, and responses within lessons for the whole class. For example, identifying some main ideas, engaging in shared reading, creating some drawings or a graphic display that illustrates their understanding, and responding by using a few words or phrases are all appropriate objectives for beginning-level English learners. In contrast, objectives for advanced students may include extensive discussion, synthesizing, reading, and report writing. At the end of this chapter, we provide an example of what this type of multilevel instruction looks like.

A key consideration for you to think about in advance of a unit of study is what you want each level of English learner to know at the end of your instruction. It is preferable to select essential content for your grade level and take the time to teach it in some depth rather than attempt to rapidly cover everything in the textbook (Chamot & O'Malley, 1994). This can provide more meaningful instruction and prevent your students from becoming overwhelmed with too much information—much of which they will not be able to understand. Stephen Covey's (1989) phrase is "begin with the end in mind." Wiggins and McTighe (1998) suggest that teachers engage in "backwards planning." Both of these phrases convey the idea that you must look over a unit of study before you teach it and make decisions about what the essential learnings are for your students to understand. What is "worthy and requiring of understanding?" (p. 64). You must answer this for each level of English learner that you are teaching, because there will necessarily be comprehension limitations connected to each level of language proficiency. A beginning-level English learner will simply not be able to understand at the same level of depth and breadth as an advanced level English learner. Once you have made essential learning decisions for your students, you will be able to make instructional decisions about how you and your students will work toward developing "enduring understandings" (p. 64).

Critical Elements of Sheltered Instruction

First, here are six research-based steps for you to follow in developing instruction:

1. Make instructional decisions about what is important for your students to learn.
2. Preview vocabulary and key concepts.
3. Pose written questions that will engage your students' interest and help focus their attention on what you want them to learn.
4. Use manipulatives, concrete objects, and realia to aid understanding.
5. Use summaries and retellings at the end of each lesson.
6. Create visual representations with your students about what they have learned.

Now we'll expand on the above by discussing in some detail four critical elements of sheltered instruction that are drawn from a program developed by the Los Angeles Unified School District (1993). These are: (1) content, (2) connections, (3) comprehensibility, and (4) interaction.

Content

There are three elements to consider with regard to the content of what you will be teaching to English learners:

- The content of what you teach is drawn from the curriculum frameworks specific to your state and district. This is the same content at the appropriate grade level that should be presented to English learners. In other words, your aim is to present the regular grade-level curriculum to English learners, although the delivery will be modified to make it accessible to them.

- Make decisions about what key learning you want your students to understand at the end of the unit of study.
- In addition to the subject matter, you will also be focusing on language development. This includes the key vocabulary relevant to the particular concepts being taught, as well as the particular language structures required to understand, speak, read, and write about the content (Diaz-Rico & Weed, 2002; Echevarria, Vogt, & Short, 2000). For example, will students be asked to describe the role of Egyptian trade in early civilizations? You may need to teach the language needed for students to be able to provide detailed descriptions, explanations, or summaries. Consider the different language levels of your students as discussed in Chapter Four and plan according to their needs. Whenever possible, we recommend that you integrate all four language skills of listening, speaking, reading, and writing, as these are closely related and can assist students with making connections between speech and print (Law & Eckes, 2000).

Connections

You can facilitate your students' learning if you make connections between new concepts and concepts they already know. New information must be linked to students' background knowledge and experiences. Here are some ways in which you can assist students to make these connections:

Activate Background Knowledge

Before teaching about a topic, encourage students to share what they know about it. One popular way to do this is to use the KWL chart. In this technique, a chart is divided into sections labeled *What We Know*, *What We Want to Know*, and *What We Learned*. Before the lesson, the students talk about what they know and want to know about a subject, and the teacher records their ideas on the chart. This activity helps to activate students' prior knowledge and stimulate their interest in the topic to be studied (Tompkins, 2006). It also provides you with essential information about what the students may already know and need to learn. After the lesson or unit of instruction, the students revisit the chart to list what they have learned. KWL charts are easy to implement, and the activity can be adapted to have students make individual charts that they share in small groups.

Connect Concepts to Students' Experiences

Focus on how the new learning is relevant to the students' lives. Student learning improves when teachers connect lessons to students' real life experiences (Moll, 1988; Vygotsky, 1978). For example, prior to a math lesson on percentages, you could show advertisements of sale items that show a third or one half off the regular price and discuss how understanding percent could save them money.

Build Background Knowledge

Provide experiences via fieldtrips, videos, demonstrations, pictures, or real objects to build background knowledge. Prior to teaching about ancient civilizations, one teacher showed students photographs of herself taken on a trip to Egypt. The fact that their teacher was in the photos made it especially interesting and appealing to the students, and it motivated them to learn more about the topic.

Comprehensibility

A key feature of effective instruction for English learners is modification of the lesson presentation so that English learners can understand important concepts being taught. This can be accomplished in a variety of different ways, some of which have been mentioned in previous chapters. It is also important to keep in mind and adapt your lessons to the different proficiency levels in your classroom. To ensure that your instruction is comprehensible, you will need to modify your speech, include many visuals to support your talk, and check frequently for comprehension. Here are some suggestions:

Teacher Talk

- For beginning English learners, articulate words clearly and slow your rate of speech, but not so much that it becomes unnatural. This will help students to be able to distinguish important words and process the language more easily (Diaz-Rico & Weed, 2002).
- Emphasize key words by repeating them frequently and using voice intonation to place more stress on these words. Repetition of key words should be naturally woven into the lesson presentation.
- Paraphrase when appropriate to clarify meaning. A teacher might say, for example, "There was a diverse group of people in the Middle Colonies in 1759. The people were from many different cultures and different religions."
- Incorporate Total Physical Response (see p. 73) with beginning students so they can acquire basic vocabulary and participate in the lesson at their comfort level. ("Point to the stem of the plant. Point to the roots.")

- Expand on students' language to model standard grammar for beginners and complex structures for more advanced students. For example, if a teacher asks, "Why did the colonists come to New England?" and a student responds, "They come for religion." The teacher might say, "Yes, they came so they could practice their religion. They came for religious freedom."

Context Clues

Use gestures, actions, real objects, pictures, or video to help students understand what you are saying. Students benefit greatly by having these additional clues to construct meaning from the lesson.

- Incorporate graphic organizers into your lessons whenever possible. These are visual displays of ideas that help students organize information so they can more easily understand and remember it. Graphic organizers help students understand, summarize, and synthesize information. They help students select important ideas and see relationships between pieces of text. Graphic organizers are powerful tools for helping English learners develop content knowledge. Graphic organizers include: KWL Charts, Venn diagrams, word webs, series of events chains, problem-solution outlines, compare/contrast matrices, network trees, cycles, and T-charts (See Appendix). *The Pictorial Input Chart* is a type of graphic organizer that offers a very appealing way to present content to students (Brechtel, 2001). First, on a large piece of butcher paper, make a drawing in light pencil of key concepts that relate to the topic of study. This is typically one picture of an object or event that captures the "big idea" you want to convey. For example, it may be the human body and the systems your students will study. As you talk through and present the initial overview of the unit, draw over your pencil lines with a marking pen. You can also write in vocabulary words. Students enjoy watching the picture emerge. This activity helps them understand the concepts and retain the information because it visually imprints it in the brain. You can repeatedly come back to these charts throughout the unit of study.

- Demonstrate to get your meaning across and to model for students. For example, as you explain the steps to solve a division problem, *model* each step of the process. *Show* how to classify leaves or rocks based on certain criteria.
- Have students dramatize the meaning of key vocabulary. In one of the classroom vignettes in Chapter Five, we illustrate how a teacher engaged her students in a brief dramatization in which students took on the roles of "waders" and a "stingray" to act out the meaning of *wading* and *prod* (Zuñiga & Yopp, 1996).
- Promote students' active participation in lessons. The more involved they are in doing something, the more likely they will understand and retain concepts. When students manipulate concrete materials, act out a social studies event, or conduct a science experiment, they improve their understanding of concepts and are more likely to remember them.

Check for Comprehension

- Monitor your students carefully to see if they are comprehending the lesson. Look for signs of frustration or behavior that indicate students may not understand and need clarification. It is not enough to simply ask if they understand, because students may not even be aware that they don't. Instead, ask questions that call for responses that would demonstrate understanding. For example, "Thumbs up if you think this is a mammal. Thumbs down if you think it is not a mammal." Asking all students to show thumbs up or thumbs down or finger signals in response to your questions are easy ways to check for whole class comprehension without singling out individual students.

- Use a variety of questions to check for understanding of concepts and adapt your questions to the language level of the students. For example, you may ask a student at a beginning level, "What are two elements in the blood?" But ask a more advanced student, "How are carbon dioxide and oxygen exchanged in the lungs and tissues?"

Interaction

We have repeatedly mentioned the importance of social interaction. Again, we stress that establishing a need to communicate with others is essential for language development. Provide your students with many opportunities to talk about what they are learning and to practice the new language in meaningful ways. As students communicate with you and with each other, both orally and in writing, about key concepts, they will improve their language skills as well as their conceptual understanding.

Cooperative Learning

A great deal has been written about the effectiveness of cooperative learning. There are many resources available to help you to implement it in your classroom (Johnson

& Johnson, 1984; Kagan, 1994). Cooperative learning is a strategy that involves students working collaboratively with partners or in small groups to achieve both academic and social goals (Peregoy & Boyle, 2005). Cooperative learning can lead to higher academic achievement, improved social relations within the class, and lowered anxiety levels for all students (Kagan, 1994). We are not advocating an exclusive use of cooperative learning methods, but rather a balance between individual and cooperative in order to prepare students for the range of learning situations they will encounter in the future.

Learning through cooperation can be especially useful for English learners because it provides excellent practice in using both social and academic English. In addition, the home cultures of some English learners may favor cooperation over competition, as is the case in traditional families of Latino and American Indian children. Thus, cooperative learning can help increase students' comfort levels. Most importantly, students can practice using academic language as they describe science experiments, compare geographical features, or explain the steps to solve a math problem. We recommend that students be grouped heterogeneously whenever possible so that students of different language and academic levels can learn from each other.

Following are three cooperative learning activities drawn from the work of Kagan (1994) that can be very useful:

- **Think-Pair-Share:** Ask a question and have students think for a couple of minutes alone, and then form pairs so students can discuss an answer with a partner. After a few minutes, call on students to share their answers or to share an answer from their partner. This works best with questions that can have more than one answer *(Name three things that . . .* or *What advice would you give . . .?)*

- **Numbered Heads Together:** Place students in small groups of four or five. Number off within groups so that each student is a 1, 2, 3, 4, or 5. (This also works well using colors). Then pose a question, such as "Make sure everyone in your group knows the reasons for . . ." Then give the students a few minutes to put their heads together to discuss the answer and make sure everyone in their group can respond. Finally, call out a number. Only the students with that number may raise their hands and respond. This motivates students to help one another while at the same time they know they will be held individually accountable.

- **Jigsaw:** Place students in "home" teams of five to seven members. Topics are assigned to home teams. One aspect of the topic or learning task is assigned to each group member, who then becomes an expert on it by working with experts from the other teams. Students meet in expert groups to learn their part of the material. Students then return to their home teams to share what they have learned and create a group project or presentation. For example, a jigsaw that focuses on the California Missions could have each home team responsible for a different mission, while expert groups study and report back to their home teams about the founding of missions, how people lived on a mission, how missions were constructed, and their present-day status. Each home team would then prepare a report on their particular mission based on a compilation of all the information from each of their experts. This cooperative learning activity can span a week or more and requires careful teacher preparation and guidance.

In addition to these cooperative learning activities, also consider these teaching tools:

- As you present information, provide frequent opportunities for students to talk about the concepts. Brechtel (2001) suggests the 10-2 strategy. After ten minutes of lecture, the teacher stops and has each student talk with a partner for two minutes about what they just learned. This allows students to practice new vocabulary and check their understanding in a comfortable environment. This is a good opportunity for students who speak the same primary language to review in that language what they've learned.
- If students are very limited in English, allow them to use their native languages to discuss concepts with other students or bilingual aids if they are available. This will assist them in developing an understanding of important concepts. As students become more proficient in English, you can encourage greater use of English to discuss what they are learning.

Reading and Writing

Your students will benefit from having numerous opportunities to engage in reading and writing as they are learning content. As students read for information, record data, or write reports, they are practicing literacy skills that are crucial for academic success. In this section we offer some guidelines and strategies for you to consider as you plan for reading and writing activities within content instruction.

Using Textbooks

Both English learners and native English speakers can encounter difficulties with content area textbooks, because they have a different structure than narrative text. For example, a common pattern for organizing information in content textbooks consists of stating a main idea and then providing a series of details to support the idea (Peregoy & Boyle, 2005). This differs from the typical structure of narrative text that tells a story and usually involves characters who encounter and solve a problem within a certain setting. In addition, textbooks are written at grade level readability, and English learners, as well as many native English-speaking students, may not be able to read at grade level. You can counter some of the difficulties that students have with content area textbooks by doing the following:

- Teach the structure of content area textbooks. Point out the particular patterns of the text they will be reading that could include cause and effect, main idea and details, or sequence of events. Use graphic organizers to help students visualize and understand the pattern. These are sometimes included in the textbooks themselves.
- Show students how to use headings and subheadings to preview text and make predictions about the content (Peregoy & Boyle, 2005). Model by "thinking aloud" how these headings can guide and generate questions for a reader.
- Show students how to use comprehension aids such as chapter overviews, graphics, and summaries (Tompkins, 2006). Point out pictures, captions, and words in bold type and help students to see how these can help them understand the text.
- Supplement the textbook with books that include some of the same content but are easier to read. Sometimes old textbooks can be useful and easier for students to read; look to see if your school has any in storage. Big books designed for upper-grade

students that present content information are also commercially available and sometimes included in English Language development programs. Some teachers have students create their own big books.

- Read the text aloud to the students. In addition to reading the text, you can pair students with reading buddies who are able to read to them. Passages from the textbook can also be tape recorded for students to listen to as they follow along in their books. Many textbook publishers now include audiotapes of their books. Parents or cross-age volunteers can also help with read-alouds.
- Prepare your own reading material. This can consist of simple paragraphs to convey key ideas. Although this can be time-consuming, it has the advantage that you can adjust the material to meet the specific needs and interests of your students.

More Meaning-Making Tools

- Preview and review lessons. Before the lesson, present your English learners with a preview of key vocabulary and a brief synopsis of the content. After the lesson, review vocabulary and concepts. This provides students with additional repetition that is extremely valuable to their learning.
- Provide students with support in their native languages whenever possible. Although a major goal with content instruction is English language acquisition, understanding the concepts being taught is just as important. Students who have limited English language skills can benefit from having either bilingual tutors or instructional assistants or books in their native languages to clarify their understanding. If you have the resources, a preview/review method can also be

done in the native language—preview the lesson in the native language and after the lesson is presented in English, provide a review in the native language.

- Use thematic instruction. It is very helpful to English learners when instruction is presented via integrated thematic units (Peregoy & Boyle, 2005). This type of instruction integrates various subject areas. For example, when learning about the Gold Rush, students can also read literature related to the period and create graphs that compare the distances people traveled. Thematic instruction supports student learning because it facilitates establishing connections and remembering concepts.
- Have students maintain *Learning Logs* in which they record what they are learning on a daily or weekly basis. Students can explain how they solved a math problem, record observations from a science experiment, or summarize what they have learned from a lesson or unit of instruction using words or drawings. You can work out a system to read and respond to students' logs so that it does not become too time consuming. Some teachers skim all logs weekly, then select four or five to respond to in writing.
- Bring in supplemental reading material from a variety of sources that relates to concepts being learned. This could include newspaper articles about current or past events, biographies, online sources, reference material, and other books (Chamot & O'Malley, 1994).
- Teach students how to take notes and develop outlines, develop graphic organizers, and write summaries of what they read. Model the procedure several times with short pieces of text, then provide students with guided practice in

small groups where they can help one another before having them do this independently.

- Have students create graphic displays that they can label with main ideas and vocabulary terms. For example, they can create time lines, illustrate the steps to solve a math problem, or conduct an experiment, design maps, develop semantic webs that include key vocabulary, or create charts to illustrate a concept.
- Have students work in small groups to develop reports and present them to their classmates. These can be oral or written reports.
- Have students write from the perspective of a historical figure. A fourth-grade teacher we know had her students write diary entries as though they were traveling to California during the Gold Rush.

Building Vocabulary

As we mentioned earlier, vocabulary development is essential for English learners' academic success. There is a strong relationship between vocabulary knowledge and academic achievement (McKeown, 1985). Without adequate vocabulary knowledge, students cannot fully understand what they read, and they will have difficulty with content instruction. Students with good vocabulary knowledge will comprehend text more easily, which leads them to more extensive reading and, in turn, greater vocabulary growth. Vocabulary development is a critically important aspect of instruction for all students and especially for English learners. Keep in mind that vocabulary is best learned within the context of interesting activities, not by memorizing lists of words or copying definitions. You will want to immerse your students in vocabulary-rich activities that promote their active involvement and provide varied opportunities for students to practice and apply knowledge of new words. Your enthusiasm for word knowledge is key to motivating students to get excited about learning new words. We discussed some ideas for vocabulary development in previous chapters. Here we present some additional suggestions to make your vocabulary instruction both engaging and challenging.

Word of the Day

This is a fun way to make vocabulary instruction part of your daily routine. Select a word that is related to a subject the students are currently studying or to an interesting upcoming event. You may also want to encourage students to volunteer a word to study. Display the word, define it, and give examples in sentences that are meaningful to the students. Engage the students in a Think-Pair-Share activity to come up with their own sentences. Finally, encourage use of the word throughout the day and place it on a word wall so that students can refer to it. You may even want to challenge students to look or listen for the word outside of class and have them share their observations.

Use Graphic Organizers

We cannot say enough about the benefits of these visuals. Not only are graphic organizers helpful for organizing information as we discussed earlier, but they assist students to picture and remember word meanings and relationships. For example, a *semantic* (or *web) map* (see Appendix) can be used in the following way (Gunning, 2003):

1. Introduce a key word and discuss it with the class.
2. Have the students come up with as many words as they can that are related to the key word and add words that you want to teach. Write all of these on the board.

3. With the students, create a semantic map on a large piece of chart paper. Guide the students to categorize the words and discuss why some words go together. The completed map will consist of the key word in the center with lines radiating out from the center that lead to related words.
4. Post the map in the classroom so students can refer to it and add more words later as they study the topic further.

Games

Students are usually motivated to learn through games. While having fun, students can also be practicing language skills. Games can be adapted to the various proficiency levels of your students and the content you are teaching. Here are a few ideas:

- **Brainstorm of Words:** Divide the class into teams of four or five students. Give a word to the class and each team has three minutes to brainstorm related words. Depending on what you are teaching, these could be antonyms, synonyms, or other words that are related to a unit of study. One student in each team can be a designated recorder to write down all the words for his/her team. After the allotted time, ask each team to share its words. Points can be given for each correct word.
- **Dramatization:** Place cards with words that students have been studying in a container and call on a student to select one. The student must pantomime the meaning of the word while the rest of the class tries to guess the word. The student who can guess the word correctly and give its meaning can be the one to select and pantomime the next word.

- **Bingo:** This is a classic game that can be adapted in many ways. Create your own Bingo cards with blank spaces. The students can then copy words that you select in random order into each of the spaces. Read aloud the definition of each word. The students then try to find a word on their card that matches the definition in order to cover it with a marker. The first one to fill a whole row of spaces calls "Bingo" and is the winner. There are many variations of this game. One is to have students find synonyms or antonyms on their cards. We once observed a third-grade teacher who had beginning-level students glue pictures from a an activity sheet onto their Bingo cards. When the teacher called out a word, the students had to match it to its corresponding picture.
- **Concentration:** This game can be played by small groups of four or five students and can be used to practice different vocabulary skills. Create two sets of cards with vocabulary words written on them. Each group places the cards facedown in rows on a table. The first student to take a turn selects two cards to turn over. If the words do not match, the cards are placed facedown in their original position. Once a student makes a match, he/she must give a definition of the word (or use it in a sentence, identify a synonym, etc). If the response is correct, the student keeps the cards. If the student is not correct, the cards are replaced again. The student with the most cards at the end wins the game.

Vignette # 7

Read the following vignette about Vance, a fifth-grade teacher. Can you identify at least six tools he uses to promote comprehension for his English learners?

In planning for the unit of study on the American Revolution, Vance decided that one concept he wanted all of his students to understand was the reason that the colonists rebelled against the British—the notion of taxation without representation. The whole class had spent a few days in advance of the unit doing a Know-Want-Learn chart and had seen a video on the American Revolution.

He gathered his beginning and intermediate English learners for a preview lesson. He asked the students to write down the words *taxes, represent,* and *representation* in their social studies journals. Then he said, "Look, here's a newspaper ad for a pair of athletic shoes. This pair is on sale for $35.99. Is that exactly what you pay—$35.99?" The students say that they would have to pay more—"Extra money," says one student. Vance nods and says that "this extra money is the tax on the shoes. It goes to the government. Taxes pay for schools, to fix the streets, and other things for us." He draws pictures as he's talking to illustrate what he's saying.

Then Vance asks the students who they elected from the class to go to student council. "Edwin," they say. "Yes, Edwin represents us—room seven. Edwin is our representative. He says what we want and what we don't want. He talks for us." Again, Vance demonstrates. Some of the students nod.

"Now we're going to do a little play with the words *taxes* and *representation,"* says Vance. He quickly moves the students around. "Jorge, you're the king of England. You sit on your special chair. Thais, you're a colonist working in your store. You're selling things in your store. Noe, you're a colonist working on your farm. You're working. You work hard. Antonio, you're the king's representative—like Edwin is our representative. Now, the king tells Antonio to get money. You go get money—taxes—from the colonists." Antonio walks from the king over to the colonists and puts his hand out as if to receive the money.

Vance says, "England needs the money. They had a big war with France for seven years." Vance demonstrates fighting and points to a picture from their social studies book. "And now they need money—taxes from the colonists. They have to pay for the war. The war costs lots of money. Colonists, do you like this? Do you want to give your money to Antonio for the king and for the government in England? Do you want to pay taxes?"

Noe bursts out, "The king—and the government in England—they tell the colonists they have to pay the money—eh—the taxes—but the colonists—we say NO! We didn't tell you we gonna pay! We no gonna pay you nothing! Go home. We don't get a . . . a . . . represent . . ."

"Yes," says Vance. "That is exactly the idea. The colonists think this is not fair. They do not want to pay taxes to England. This is *taxation with no representation*. The colonists did not agree to pay extra taxes. The government in England did not ask them. They just made them pay. Now you can write this important idea in your journals. The colonists did not want to pay taxes to England. It was taxation with no representation."

Tools

What tools does the teacher use?

1. engages in backwards planning
2. selects a few key words and ideas to convey to his students in advance of the unit
3. uses a KWL chart and a video to activate vocabulary and background knowledge
4. makes personal connections to the student's lives—paying taxes for purchases and a classmate who is a representative in the student council
5. repeats and emphasizes key ideas and words
6. incorporates TPR (i.e., tells Antonio—"You go get money—taxes—from the colonists," and Antonio then puts out his hand to receive the money.)
7. draws pictures as he talks
8. has students do a simple dramatization to highlight key words
9. has students write down key concepts and vocabulary in a content journal—makes listening and writing connections

Sample Content Lesson (Social Studies)

Following is a sample content lesson from the fourth-grade California social studies curriculum on the Gold Rush. We've included the California history/social science standard and the state English Language Development (ELD) standards for each of the five levels of English language proficiency. The sample lesson illustrates multilevel objectives, how you might group students for effective instruction, and teaching tools you can use that are effective for the various levels of English language development.

Grade: 4th

Content Area: California History/Social Science Content Standard 4.3.3.

Analyze the effects of the Gold Rush on settlements, daily life, politics, and the physical environment.

State ELD Standards:

Beginning:

- answer simple questions with one- or two-word responses
- retell simple stories using drawings, words, or phrases

Early Intermediate:

- orally identify the main points of simple conversations and stories that are read aloud using phrases or simple sentences
- produce independent writing that is understood when read but may include inconsistent use of standard grammatical forms

Intermediate:

- use content-related vocabulary in discussions and reading

- use more complex vocabulary and sentences appropriate for language arts and other content areas
- produce independent writing that is understood when read but may include inconsistent use of standard grammatical forms

Early Advanced:

- retell stories in greater detail including characters, setting, plot summary, and analysis
- ask and answer instructional questions with more extensive supporting elements
- write multiparagraph narrative and expository compositions and examples appropriate for content areas, with consistent use of standard grammatical forms

Advanced:

- speak clearly and comprehensively using standard English grammatical forms, sounds, intonation, pitch, and modulation
- read narrative and expository text aloud with appropriate pacing, intonation, and expression
- create multi-paragraph narrative and expository compositions using standard grammatical forms

Lesson's key concept: The gold found at Sutter's Mill in 1848 caused many people to come to California.

Objectives

Beginning Level: By creating a pictorial time line and labeling it with appropriate words and phrases, students will demonstrate understanding of some main ideas and vocabulary words related to the Gold Rush.

Early Intermediate Level: Students will use sentence frames to retell the main events related to the Gold Rush.

Intermediate Level: Using vocabulary from the lesson and structured paragraphs, students will explain why the Gold Rush became important to California.

Early Advanced and Advanced Level English Learners and Native English Speakers: Using descriptive vocabulary and supporting details, students will write a multi-paragraph essay that describes the Gold Rush and explains why it became important to California and the world. (**Note**: Early advanced level students' use of standard grammatical form may be less consistent than for advanced students.)

Materials:

- small pieces of rock, painted gold
- photocopied pictures and subheadings from a social studies book
- sieve
- "newspaper" with headline "Gold in California!"
- a map of the United States
- a time line

Vocabulary:

- gold nugget
- discover/discovery
- miner
- forty-niners
- journey
- pan for gold
- Gold Rush
- gold fever

Prior to the Lesson:

Divide the class into two groups. Native English speakers and advanced level ELD students will work in pairs to complete the activity sheet below to help build interest and background knowledge. While they are completing the activity sheet, meet with the beginning through early advanced level English learners to introduce key vocabulary and concepts using pictures and props.

> **Directions:** With your partner, preview the chapter on the Gold Rush in your social studies book. Look carefully at the pictures and read the subtitles. Then talk about and complete this assignment together.
>
> 1. Why do you think the discovery of gold caused a "rush"?
> 2. Why did we call it the *Gold Rush*?
> 3. How did gold end up in the American River?
> 4. What do you suppose would be three challenges gold seekers would encounter?
> 5. How do you think the Gold Rush changed California? Write one question to which you'd like to know the answer.
> 6. List three words you don't know in this chapter. Later, we'll talk about your "mystery words."

While others are working on the above questions, meet with beginning through early advanced level English learners to introduce key vocabulary and concepts using pictures and props.

Lesson Development

Teacher modeling/input

Bring the whole class together. Using pictures and props, describe the discovery of gold at Sutter's Mill. Show text pictures of miners panning for gold and use gold-painted rocks in a sieve to demonstrate "panning for gold." Show and discuss the "newspaper" with the important discovery in California, a map of the United States, text pictures with various modes of arrival to California, and pictures of a Forty-niner and some of his equipment. At appropriate points, ask advanced English learners and native English speakers to share their responses to preview questions. Ask them to share their "mystery words" and discuss their meaning.

Guided practice

As the material is presented, ask the students a variety of questions such as: *Who can point to the gold nugget? Who can show me "panning for gold"? How did people get to California? Was it easy or difficult? Why did the Forty-niner come to California? How did he get here? What is he doing? What does he want? Did other people come? How many? What do you think happened when so many people arrived in California?* Write some of the students' responses that include key vocabulary items.

For beginning and early intermediate students: Give the students the time line and the photocopied pictures and subheadings from the text. Ask students to select and paste them in order and to write words and phrases on the time line that explain the series of events.

Intermediate and early advanced students: Have students create a graphic organizer with key words and use it to write one or more structured paragraphs describing the main events and their significance.

Advanced and native English speakers: Have students create a graphic organizer with key ideas and use it to write several paragraphs that describe the main events and their significance.

Bring the whole class together again. Have students share and discuss their work as a whole class. Ask one or two students to share their time lines. An intermediate or early advanced English learner can describe the time line, if needed. Ask one or two students to read a structured paragraph. Ask advanced English learners and native English speakers to share one or more of their paragraphs.

Closure: Ask students to write important words and ideas from today's lesson in their social studies journals.

Important Points to Remember

- Make connections—connect new concepts to previous learning and relate it to students' real life experiences whenever possible.
- Provide comprehensible input—use a variety of scaffolding techniques to make your instruction as comprehensible as possible. This includes modifying your teacher talk and supporting it with many visual clues.
- Promote interaction—provide many opportunities for students to work with partners or in small groups to practice social and academic language skills.
- Take an active interest in your students and learn as much as you can about their backgrounds and interests. The more you know about them, the more you will be able to plan for instruction that builds on their background knowledge.
- Maintain high expectations for all of your students. When you establish a climate that says you expect students to perform well, they usually do (Henze & Lucas, 1993).

Post-Reading Reflection

1. What is "sheltered instruction" and what is its purpose?

____

____

____

2. Identify two ways that you could help English learners to make meaningful connections between new concepts and concepts they already know.

____

____

____

3. List three ways you could modify content instruction to make it more comprehensible for English learners.

____

____

____

4. What are two ways you can provide opportunities for students to practice their language skills and review the concepts they are learning?

____

____

____

References

Adams, M. J. (1990). *Beginning to read: Thinking and learning about print.* Cambridge, MA: MIT Press.

Asher, J. (1982). *Learning another language through actions: The complete teacher's guide book* (2nd ed.). Los Gatos, CA: Sky Oaks Productions.

Austin, J.L. (1962). *How to do things with words.* Oxford, England: Oxford University Press.

Benson, V., & and Cummins, C. (2000). *The Power of Retellings.* Bothell, WA: Wright Group/McGraw-Hill.

Brechtel, M. (2001). *Bringing it all together: Language and literacy in the multilingual classroom.* Carlsbad, CA: Dominie Press.

Bromley, K. (1995a). Buddy journals in the classroom. In N. Hall, & A. Robinson, (Eds.), *Keeping in touch: Using interactive writing with young children* (pp. 42–54). Portsmouth, NH: Heineman.

Bromley, K. (1995b). Buddy journals for ESL and native-English-speaking students. *TESOL Journal (Spring),* 7–11.

California Department of Education (1996). *Practical ideas for teaching writing as a process at the elementary and middle school levels.* Sacramento, CA: California Department of Education.

Callahan, R.M. (2005). Tracking and high school English learners: Limiting opportunities to learn. *American Educational Research Journal, 42(2*), 305–328.

Chamot, A.U., & O'Malley, J.M. (1994). *The CALLA Handbook: Implementing the cognitive academic language learning approach.* New York: Longman.

Chiape, P., Siegel, L. S., & Wade-Woolley, L. (2002). Linguistic diversity and the development of reading skills: A longitudinal study. *Scientific Studies of Reading, 6(4)*, 369–400.

Chomsky, N. (1965). *Aspects of the theory of syntax.* Cambridge, MA: MIT Press.

Collier, V. P. (1987). Age and rate of acquisition of second language for academic purposes. *TESOL Quarterly, 21,* 617–641.

Collier, V. P. (1989). How long? A synthesis of research on academic achievement in second language. *TESOL Quarterly, 23,* 509–539.

Collins, J.L. (1998). *Strategies for struggling writers.* New York: Guilford.

Covey, S.R. (1989). *Seven habits of highly effective people.* New York: Fireside.

Crawford, A. N. (1994). Communicative approaches to second language acquisition: From oral language development into the core curriculum and L2 literacy. In California State Department of Education (Ed.), *Schooling and language minority students: A theoretical framework* (2nd ed., pp. 79–131). Los Angeles: Evaluation, Dissemination and Assessment Center, California State University.

Cummins, J. (1981). The role of primary language development in promoting educational success for language minority students. In California State Department of Education (Ed.), *Schooling and language minority students: A theoretical framework* (pp. 3–49). Los Angeles: Evaluation, Dissemination and Assessment Center, California State University.

Cummins, J. (1994). Primary language instruction and the education of language minority students. In California State Department of Education (Ed.), *Schooling and language minority students: A theoretical framework* (2nd ed., pp. 3–46). Los Angeles: Evaluation, Dissemination and Assessment Center, California State University.

Cummins, J. (1996). *Negotiating identities: Education for empowerment in a diverse society.* Los Angeles: California Association for Bilingual Education.

Cunningham, P. M. (2000). *Phonics they use: Words for reading and writing* (3rd ed.). New York: Harper Collins.

De Jong, E., & Harper, C., (2005). Preparing mainstream teachers for English-language learners. *Teacher Education Quarterly, 32(2),* 101–124.

Diaz-Rico, L.T. (2004). *Teaching English learners: Strategies and methods.* Boston: Pearson.

Diaz-Rico, L.T., & Weed, K. Z. (2002). *The crosscultural, language and academic development handbook* (2nd ed.). Boston: Allyn & Bacon.

Dunlap, C. Z. & Weisman, E. M. (2005, April). *Vietnamese-English and Spanish-English preservice teachers: A comparison of perceptions of the bilingual teacher.* Paper presented at the annual meeting of the American Educational Research Association, Montreal, Canada.

Echevarria, J., Vogt, M., & Short, D. J. (2000). *Making content comprehensible for English learners.* Boston: Allyn & Bacon.

Fillmore, L.W., & Valdez, C. (1986). Teaching bilingual learners. In M.E. Wittrock (Ed.), *Handbook of research on teaching* (pp. 648–685). New York: Macmillan.

Fitzgerald, J., & Noblitt, G.W. (1999). About hopes, aspirations, and uncertainty: First grade English-language learners' emergent reading. *Journal of Literacy Research, 31(2),* 133–182.

Gandara, P., Maxwell-Jolly, J., & Driscoll, A. (2005). *Listening to teachers of English learners.* Santa Cruz, CA: Center for the Future of Teaching and Learning. Retrieved July 17, 2005, from http://www.cftl.org/publications_latest.php.

Goodman, K. S. (1996). *On reading.* Portsmouth, N.H: Heinemann.

Gunning, T. G. (2003). *Creating literacy instruction for all children* (4th ed.). Boston: Allyn & Bacon.

Hatch, E. (1992). *Discourse and language education.* Cambridge, England: Cambridge University Press.

Henze, R. C., & Lucas, T. (1993). Shaping instruction to promote the success of language minority students: An analysis of four high schools classes. *Peabody Journal of Education: Trends in Bilingual Education at the Secondary Level, 69(1),* 54–81.

Johnson, R.T., & Johnson, D.W. (1984). *Structuring cooperative learning: Lesson plans for teachers.* Minneapolis: MI: Interaction Book Company.

Kagan, S. (1994). *Cooperative learning.* San Juan Capistrano, CA: Resources for Teachers.

Kalmbach, J.R. (1986). Getting the point of retellings. *Journal of Reading, 29,* 326–333.

Koch, K. (1970). *Wish, lies, and dreams: Teaching children to write poetry.* New York: Perennial Library.

Koskinian, P., Gambrell, L., Kapinus, B., & Heathington, B. (1988). Retelling: A strategy for enhancing students' reading comprehension. *The Reading Teacher, 41*, 892–96.

Krashen, S. (1982). *Principles and practices in second language acquisition.* Oxford, England: Pergamon Press.

Krashen, S. (1996). *Under attack: The case against bilingual education.* Culver City, CA: Language Education Associates.

Krashen, S. D., & Terrell, T. T. (1983). *The natural approach: Language acquisition in the classroom.* Oxford, England: Pergamon Press.

Law, B., & Eckes, M., (1995). *Assessment and ESL.* Winnepeg, Canada: Peguis.

Law, B., & Eckes. M. (2000). *The more than just surviving handbook* (2nd ed.). Winnipeg, Canada: Portage & Main Press.

Lessow-Hurley, J. (1999). *The foundations of dual language instruction* (3rd ed.). Boston: Allyn & Bacon.

Lightbown, P. M., & Spada, N. (1999). *How languages are learned.* Oxford, England: Oxford University Press.

Long, M. H. (1983). Native-speaker/non-native speaker conversation and the negotiation of comprehensible input. *Applied Linguistics, 4*, 126–41.

Los Angeles Unified School District (1993). *Sheltered instruction teacher handbook: Strategies for teaching LEP students in the elementary grades.* Los Angeles: LAUSD.

McKeown, M. G. (1985). The acquisition of word meaning from context by children of high and low ability. *Reading Research Quarterly, 20*, 482–496.

Meskill, C. (2005). Infusing English language learner issues throughout professional education curricula: Training all teacher project. *Teachers College Record, 4*, 739–756.

Minneapolis Public Schools (2005). *Facts you should know—student demographics.* Retrieved July 17, 2005, from http://www.mpls.k12.mn.us/MPS_Facts.html#demographics.

Moll, L. (1988). Some key issues in teaching Latino students. *Language Arts, 65,* 465–472.

Morrow, L. M. (1993). Using story retelling to develop comprehension. In K. Muth (Ed.) *Children's comprehension of text: Research into practice* (pp. 24–36). Newark, DE: International Reading Association.

Nieto, S. (2004). *Affirming diversity* (4th ed.). Boston: Pearson Education.

Ogulnick, K., Shelton-Colangelo, S., & Williams, C., (1998). Entering the fictive world: Enhancing the reading experience. *Cultural Circles, 3,* 125–133.

O'Malley, J.M., & Valdez Pierce, L. (1996). *Authentic assessment for English language learners: Practical approaches for teachers.* New York: Addison-Wesley Publishing.

Parker, F., & Riley, K. (2000). *Linguistics for non-linguists.* (3rd ed.). Boston: Allyn & Bacon.

Peregoy, S. F., & Boyle, O. F. (2000). English learners reading English: What we know, what we need to know. *Theory into Practice, (39)4,* 237–247.

Peregoy, S. F., & Boyle, O. F. (2005). *Reading, writing, and learning in ESL: A resource book for K–12 teachers* (4th ed.). Boston: Pearson Education.

Piper, T. (1998). *Language and Learning.* (2nd ed.). Upper Saddle River, NJ: Merrill.

Ramirez, J.D. (1992). Executive summary of the final report: Longitudinal study of structured English-immersion strategy, early-exit and late-exit transitional bilingual programs for language-minority students. *Bilingual Research Journal, 16,* 1–62.

Ramirez, D., Yuen, S., & Ramsey, D. (1991). *Final report: Longitudinal study of structured English immersion strategy, early-exit, and late-exit transitional bilingual education programs for language minority children.* Washington, DC: Office of Bilingual Education.

Ruddell, R. (2006). *Teaching children to read and write.* Boston: Pearson, Allyn & Bacon.

Schiffini, A. (1996). Reading instruction for the preliterate and struggling older students. *NABE News, 20,* 5–6, 20, 30.

Temple, C., Ogle, D., Crawford, A., & Freppon, P. (2005). *All children read.* Boston: Allyn & Bacon..

Terrell, T. D. (1981). The natural approach in bilingual education. In California State Department of Education (Ed.), *Schooling and language minority students: A theoretical framework* (pp. 117–146). Los Angeles: Evaluation, Dissemination and Assessment Center, California State University.

Thonis, E. (1994) Reading instruction for language minority students. In California State Department of Education (Ed.), *Schooling and language minority students: A theoretical framework* (2nd ed., pp. 165–202). Los Angeles: Evaluation, Dissemination and Assessment Center, California State University.

Tompkins, G. (2006). *Literacy for the 21st century: A balanced approach* (4th ed.). Upper Saddle River, N.J: Pearson Education.

Vygotsky, L. S. (1978). *Mind in society*. Cambridge, MA: Harvard University Press.

Wiggins, G., & McTighe, J. (1998). *Understanding by design.* Alexandria, VA: Association for Supervision and Curriculum Development.

Yopp, H.K., & Yopp, R.H. (2000). Supporting phonemic awareness development in the classroom. *The Reading Teacher, 54(2)*, 130–143.

Yopp, H.K., & Yopp, R.H. (2001). *Literature-based reading activities* (3rd ed). Boston: Allyn & Bacon.

Zuñiga, C., & Yopp, R. H. (1996). Practices of exemplary elementary school teachers of second language learners. *Teacher Education Quarterly, 23(1)*, 83–97.

Appendix

Compare/Contrast Matrix

Use to categorize, compare and process information. Can be used with any content area, including literature (compare characters, settings, plots, etc.).

Animal	Habitat	Food	Natural Enemies

Cycles

Use to illustrate events or phenomena that occur in a continuous cycle.

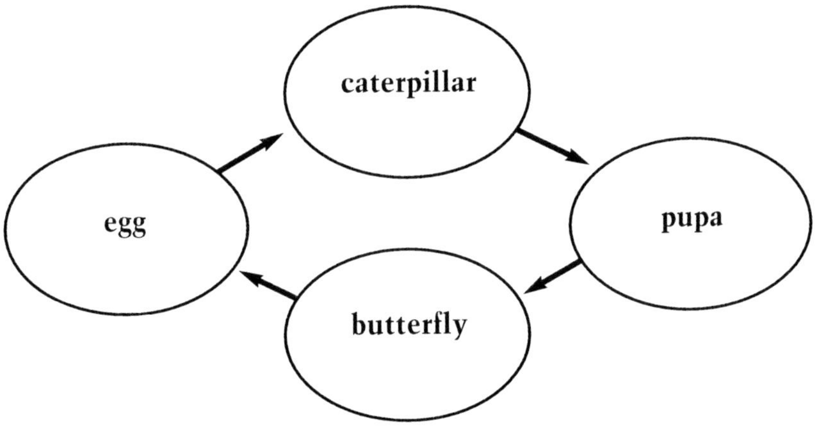

KWL Chart

Use to activate students' background knowledge and identify questions they have about the topic. At the end of the unit, students list what they learned.

What I Know	What I Want to Know	What I Learned

Network Tree

Use to illustrate the categories and examples of a main topic or concept. Write the main topic on the tree trunk, the major categories on the branches, and the subcategories or examples on the smaller branches.

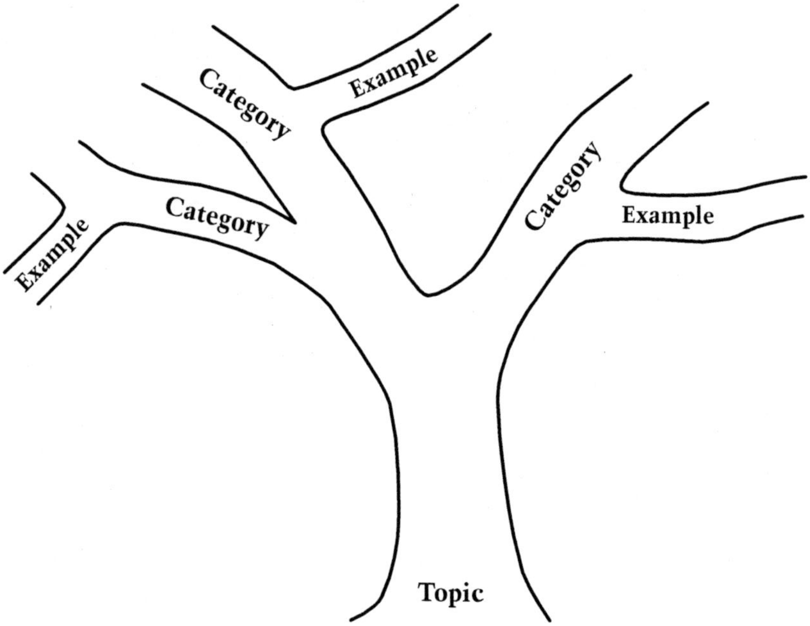

Problem-Solution Outline

Use to show a problem, possible solutions, and likely outcomes or results of proposed solutions.

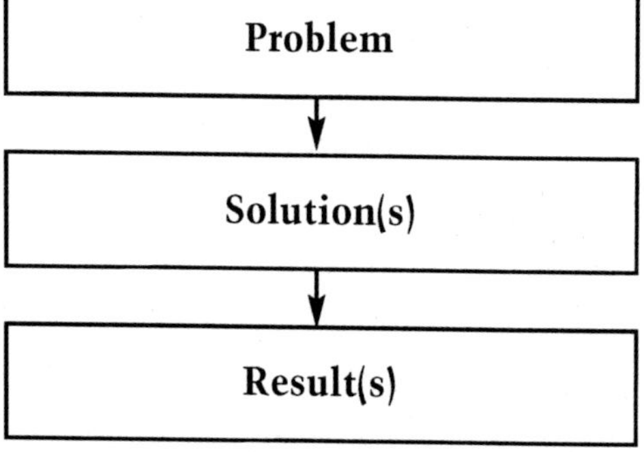

Semantic (or Web) Map

Use to map out the main ideas and details of a topic or concept. You can also use it to teach word meanings and relationships. This graphic is sometimes referred to as a **Word Web.** Here are two examples.

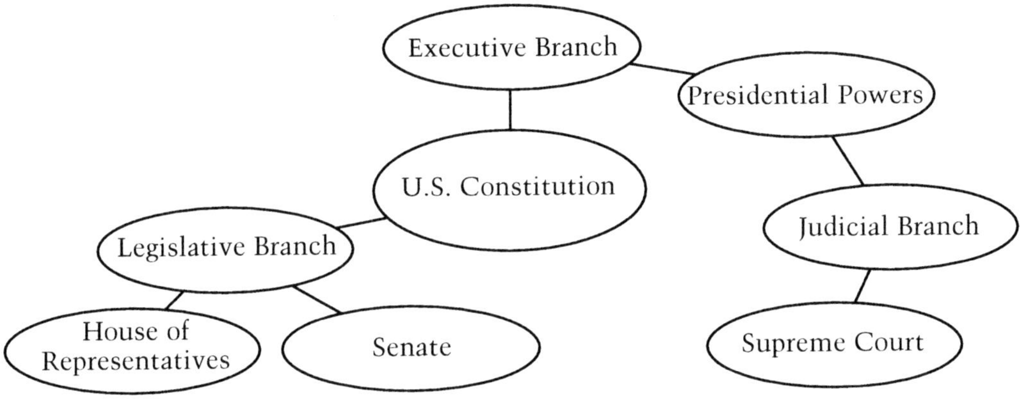

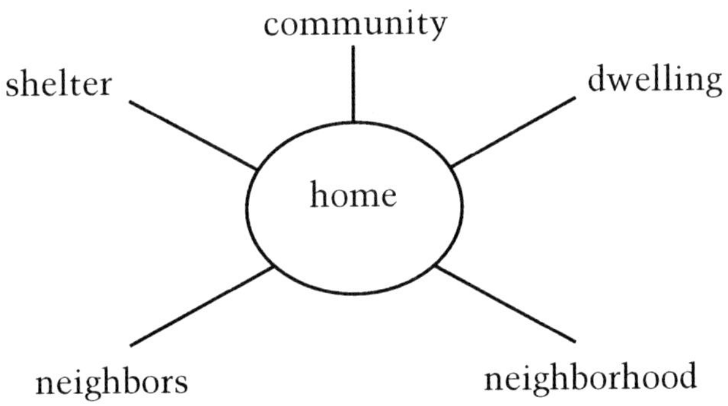

Series-of-Events Chain

Use to show sequence of events, stages of a life cycle or goals, and actions and outcomes of a historical figure or character in a novel.

T-Chart

Use to compare two things (book characters, events, ideas, etc.).

Planes	**Cars**

Venn Diagram

Use to compare and contrast two things (i.e., books, movies, characters, etc.).

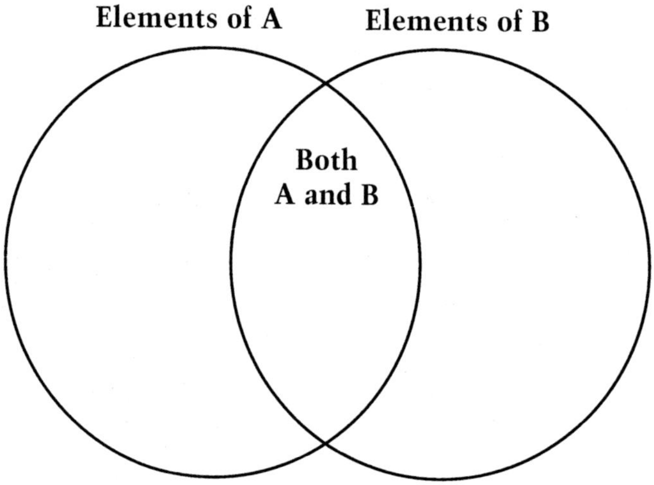

Final Words

We recognize that teaching English learners presents challenges but also offers many rewards. We applaud your efforts to improve your knowledge about teaching your English learners. As their numbers increase in your schools, your knowledge, skills, and abilities will become more valuable.

We hope you have found this book useful, and that you will be able to put its content to good use.

Best wishes for successful teaching!

Carmen and Evelyn

Notes

Notes

Notes